China's Big Science Facilities

"Big science" facilities are major elements of science and technology infrastructure, and important symbols of China's scientific and technological development. This popular science book series presents the background, history and achievements of the Chinese Academy of Sciences in terms of constructing and operating big scientific facilities over the past few decades.

The series highlights the major scientific facilities constructed in China for pioneering research in science and technology, and uses straightforward language to describe the facilities, e.g. the fully superconducting Tokamak fusion test device (EAST), the National Protein Science Research Facility, Lanzhou Heavy Ion Accelerator, Five-hundred-meter Aperture Spherical Telescope (FAST), etc. It addresses the respective facilities' research fields, scientific backgrounds, technological achievements, and strategic and fundamental contributions to science, while also discussing how they will improve the development of the national economy. Supplementing the main text with a wealth of images and linked videos, the book offers extensive information for members of the general public who are interested in scientific facilities and related technologies.

More information about this series at http://www.springer.com/series/16530

Baonian Wan
Editor

Man-Made Sun

Experimental Advanced Superconducting
Tokamak (EAST) Fusion Reactor

Editor
Baonian Wan
Institute of Plasma Physics
Chinese Academy of Sciences
Hefei, Anhui, China

Translated by
Xiaodong Chen
Nanjing Normal University
Nanjing, Jiangsu, China

ISSN 2662-768X   ISSN 2662-7698   (electronic)
China's Big Science Facilities
ISBN 978-981-16-3889-3   ISBN 978-981-16-3887-9   (eBook)
https://doi.org/10.1007/978-981-16-3887-9

Jointly published with Zhejiang Education Publishing House
The printed edition is not for sale in China Mainland. Customers from China Mainland please order the print book from: Zhejiang Education Publishing House.

Translation from the Chinese language edition: 人造太阳——*EAST*全超导托卡马克核聚变实验装置 by Baonian Wan, and Xiaodong Chen, © Zhejiang Education Publishing Group 2017. Published by Zhejiang Education Publishing Group. All Rights Reserved.
© Zhejiang Education Publishing House 2021

This Springer imprint is published by the registered company Springer Nature Singapore Pte Ltd.
The registered company address is: 152 Beach Road, #21-01/04 Gateway East, Singapore 189721, Singapore

Editorial Board

Editor-in-Chief

Baonian Wan

Editorial Board Members

Xiaodong Zhang
Yuntao Song
Peng Fu
Xinchao Wu
Youzhen He
Junling Chen
Junjun Zhao
Guosheng Xu

Editors

Suzhen Zhang
Teng Wang
Rong Yu
Yaqin Li
Jie Huang
Wei Peng
Hualong Ye
Xiaofeng Han

Series Foreword

As a new round of technological revolution is bourgeoning, it will exert a direct impact on survival of a country whether or not it can gain insight on the future technological trends and grasp new opportunities from the revolution. In face of the major opportunities in the twenty-first century, China is intensively formulating the innovation-driven development strategy and building an innovation-based country in this critical era to achieve a moderately prosperous society in an all-round way.

Scientific and technological innovation and popularization remain two wings for innovation-driven development of a nation. In particular, popular science affects the awareness of the general public for science and technology as well as social and economic development. Scientific education is thus highly practical for implementing the innovation-driven strategy. Contemporary science pays more attention to public experience and engagement. The word "public" covers various social groups that exclude those in scientific research institutions and departments. The "public" also includes decision-makers and management personnel in government agencies and enterprises, media workers, entrepreneurs, science and technology adopters and others. Barriers that impede the innovation-driven strategy will emerge if any of the group fall behind this new revolution; avoiding and removing the possible barriers will strategically improve the quality of human resources, enhance mass entrepreneurship and innovation and build a moderately prosperous society in an all-round way.

Science workers are primary creators of scientific knowledge who undertake the mission and responsibility for science popularization. As a national strategic power in science and technology, Chinese Academy of Sciences (CAS) has always shouldered and attached equal importance to this mission in addition to scientific innovation and incorporated the mission into key measures of the "Pioneering Action" plan. CAS enjoys rich and high-end technological resources, such as the high-caliber experts represented by CAS members, advanced research facilities and achievements represented by the Big Science Project, excellent scientific popularization base represented by the national scientific research and popularization base. With these resources in place, CAS implements the "High-level Scientific Resource Popularization" plan to transform the resources into popular facilities, products and talents to benefit trillions of public. Meanwhile, CAS launches the "Science and China" program, a scientific education plan, to mobilize more effectively the "popularized high-end scientific research resources" for scientific education targeted at the public and the integration of science and education.

Scientific education requires not only dissemination of scientific knowledge, approaches and spirit to enhance overall scientific literacy of the country, but also creation of scientific environment to enable scientific innovation to lead sustainable and sound social development. For this reason, CAS cooperates with Zhejiang Education Publishing House to launch the CAS Scientific and Cultural Project. This is a large-scale scientific and cultural communication project on the basis of CAS research findings and expert team to improve the scientific and cultural quality of the Chinese citizen in an all-round manner and to serve for the strategy of the national rejuvenation by advancing science and technology. On the basis of the target group, the project is categorized to two series, i.e., the Adolescent Scientific Education and the Public Scientific Awareness, respectively, for the adolescent and the general public.

The Adolescent Scientific Education series aims to create a series of publications that draw on frontier scientific research findings and introduce the status quo of scientific development in China; to cultivate the adolescent's interest in science study; to educate them about basic scientific research approaches; and to inspire them to develop reason-based scientific way of thinking.

The Public Scientific Awareness series aims to educate the general public about basic scientific approaches and the social significance of science and encourage the public to engage in scientific affairs, thus the project will enhance the capacity the public of conscientiously applying science to their life and production activities improve efficiency and promote social harmony. In the near future, publication series of CAS Scientific and Cultural Project will constantly come out. I hope that these publications will be welcomed by the reader and that through coordination among CAS science workers, science icons such as Qian Xuesen, Hua Luogen, Chen Jingrun and Jiang Zhuying will be more familiar to the public. As a result, the truth-pursuing spirit, reason-guided thinking and scientific ethics will be fully promoted, and the spirit of science workers in courageous exploration and innovation stands eternally in the history of human civilization.

July 2016

Chunli Bai
President of Chinese Academy of Sciences
Secretary of Leading Party Members' Group

Preface

Energy is a driving source for the development of human society.

Fossil fuels include coal, oil and natural gas and constitute the primary energy in global energy mix. As precious wealth preserved over hundreds of millions of years in nature, fossil fuels have been exploited since the Industrial Revolution in the mid-eighteenth century and have created unprecedented social prosperity for human beings. Nevertheless, the reserve of the non-renewable fossil fuels is limited. This means the fuel will be ultimately run out one day in the future. In this case, how should humanity sustain the huge demand for energy?

Nuclear fusion has the advantages of high-density, abundant raw material reserves, safe and clean usage. It thus presents an ideal major energy option in the future. According to estimates, the energy generated in all fusion reactions of deuterium contained in 1 l seawater is equivalent to the energy generated in full combustion of 300 l gasoline. Deuterium preserved in ocean is available for human beings in billions of years, and the by-products most easily generated by deuterium fusion reactions are helium and neutron. Helium is clean and safe; neutron has extensive usage in addition to the energy stored within it. These advantages make nuclear fusion a strategic energy. Countries around the world, especially developed ones, are relentlessly conducting controlled nuclear fusion research for this purpose.

The energy released by the sun mainly comes from nuclear fusion reactions. Controlled nuclear fusion follows the same principle as the sun, which is why nuclear fusion is also called the artificial sun. Magnetic confinement is one way to achieve controlled fusion. A certain amount of nuclear fusion fuels is heated in various means to a high temperature of 100 million degrees Celsius. Under such high temperature, atoms turn completely ionized and form—together with electrons—plasmas or the so-called the fourth state of matter. Confined by magnetic fields, charged particles can only move along the magnetic induction line. This feature enables setup of a reaction chamber with magnetic condenser to ensure no contact of the high-temperature fusion fuel with the chamber wall and finally achieve controlled nuclear fusion reactions.

Tokamak is an experimental facility developed by humanity to explore nuclear fusion energy. The idea was initially raised by Russian scientists. Tokamak is a form of magnetic confinement. In order to obtain a more stable confinement magnetic field, scientists utilized superconducting materials to produce tokamak fields which connect to liquid helium. Through the field, tokamak cools down to an extremely low temperature to achieve superconductivity and stable operation of the facility. In the future reactors, plasma discharge in hundreds or thousands of amperes along with toroidal magnetic fields in tens of thousands of Gauss will form spiral magnetic induction lines which twine on the tyro-shaped torus. This configuration helps avoid direct loss of combustion particles along the magnetic induction line, reduce drift loss and prevent the high-temperature plasma from directly contacting the chamber wall through configuration control.

The Experimental Advanced Superconducting Tokamak (EAST) is located in the Science Island of the west suburb of Hefei City. The facility is supported by the Big Science Project under the "Ninth Five-Year Plan" in China. Independently designed and developed by China, it is the first superconducting, non-circular, cross-sectional nuclear fusion tokamak experimental facility successfully developed and operated. EAST is also one of the experimental platforms in the world that enables research on long-pulse fusion plasma physics and engineering technologies with high parameters. Meanwhile, EAST is a nuclear fusion experimental platform and research center open not only to China but also to the rest of the world. EAST has provided guidance for future design and operation of China's fusion projects. What's more, its research findings have laid a pivotal foundation in engineering technologies and physics for building more stable, efficient and secure tokamak fusion reactors.

This book mainly introduces research background and fundamentals of magnetic confinement fusion and the process of magnetic fusion research. Written in popular language, it tries to educate the public about knowledge on what is artificial sun, why we create it and how to realize this vision step by step. We hope that this book can serve as a readable option for popular science. This book is jointly edited by popular science volunteers from the Institute of Plasma Physics, Chinese Academy of Sciences (IPP, CAS). We feel grateful to the editors of Zhejiang Education Publishing House for valuable advice and to all colleagues for planning, editing, proofreading and publishing this book. We hope that readers can have a full understanding of controlled fusion through this book. Controlled nuclear fusion is a research field that requires generations of continuous hard work. This book will be more meaningful if it attracts young readers to embark on China's controlled fusion research.

July 2017

Baonian Wan
Director
Institute of Plasma Physics
Chinese Academy of Sciences
Hefei, China

EAST is the only operating tokamak in the world that is full installed with superconducting magnets and similar to the International Thermonuclear Experimental Reactor (ITER). Despite limitation of fund, EAST was built, developed and put into first operation very quickly, a remarkable achievement that manifests the capacity of the CAS physics and engineering research team.

Complete design, pre-research, development and operation in such short period of time create a remarkable achievement in the world's fusion project. They also serve as an explicit forecast of what China will contribute to ITER in the future. This brilliant success is a historic milestone in nuclear fusion energy development of the whole world.

—EAST International Consulting Committee

China made fusion history.

—Review article from Nature

This is where the value of science is brilliantly displayed.

—Review article from Science

Contents

The Conquest of the Energy Kingdom

Suzhen Zhang

Abstract What is energy? What are energy resources? What does energy look like? Where does energy in the universe come from? Where does energy of the sun which lights up the earth and nurtures living beings come from?

1 Energy Is All Around

What is energy? Physicists will tell you that energy is the dynamic form of matter. To put it simple, energy and matter are equivalent and mutually converted. Energy is the matter released and matter is the concentrated energy. If the world is made of matter, it is also made of energy. This will be more easily understood if you know about Einstein's mass-energy equation—$E = mc^2$.

What is energy? What does it look like? These are hard to explain. Energy is like Sun Wukong in *the Journey to the West* who masters 72 earthly transformations. Energy can turn into heat, sound, or vibration. Energy is everywhere; it's everything in the world.

Energy is omnipotent; without energy everything is impossible. Energy is closely related to social progress of humanity. Activities of the human body and manual labor both consume energy. Energy plays a critical role in everyday life. Without energy, nothing will be possible for humanity.

Energy is very friendly. We can use energy to boil water, cook meal and drive vehicles; or we can use it to transmit electricity to power TVs, computers and other home appliances. Humanity can't live without energy. However, energy sometimes is very unfriendly. It also induces earthquakes, floods and tropical storms which can destroy family of humanity (Figs. 1 and 2).

How energy comes about is a mystery. How does energy in the universe originate? There is no certain answer yet. Most physicists believe that all energy in the universe came from a Big Bang in which the universe was born. Energy maybe the residues formed after the opposite matters were annihilated or the byproduct from negative

S. Zhang (✉)
Institute of Plasma Physics, Chinese Academy of Sciences, Hefei, Anhui, China
e-mail: yihco0101@sina.com

© Zhejiang Education Publishing House 2021
B. Wan (ed.), *Man-Made Sun*, China's Big Science Facilities,
https://doi.org/10.1007/978-981-16-3887-9_1

1

Fig. 1 Humanity can't live without energy

energy of gravitation. Anyway, energy has existed before we know about the universe (Fig. 3).

Knowledge Link

Matter and Energy, Which Came About First?

Matter and energy, which came about first? A probe into the universe reveals that energy obviously came before matter, because the later didn't even exist when the Big Bang occurred. And even electrons, protons and neutrons which constitute atoms did not exist at that time. These all came into being after the Big Bang according to the equation $m = E/c^2$. In this sense, energy is more fundamental than mass.

Fig. 2 Energy also leads to disasters

Fig. 3 Energy has existed since the universe began

Fig. 4 Energy resources offer humanity energy

For most of us, how energy came about is not important. What's important is how to obtain energy for our daily use. Therefore, humanity constantly search for matter that can offer us energy such as water, wind and oil. The matter from which we gain energy is called energy resources which mean "the source of energy" (Figs. 4 and 5).

There are three forms of energy on earth. The first is from outside earth, mainly the sun; the second is from inside earth, including thermal energy and nuclear energy; the third is from interaction between earth and other planets, exemplified by tidal energy.

2 All Living Things on Earth Live on the Sun

The sun has been chanted in verses, ditties, odes and songs—the four forms of poetry since ancient times. Despite a very ordinary star, the sun is our center compared with

Fig. 5 Energy is transmitted through grid

the smaller earth, because "all things on earth live on sun" and we will not longer survive without it.

The sun we see with our eyes is actually the photosphere. Temperature of this relatively "cool" surface is 6000 °C, while that of the core reaches 15 million degrees Celsius. Energy of the sun mainly comes from its core. The extremely high temperature makes the sun a huge "fireball" radiating tremendous energy all around (Fig. 6).

The heat given out by the sun is equivalent to the energy generated from combustion of 10,000 trillion tons of coal. Yet only 1/2.2 billion of the energy is received by the earth. Such small portion equals, however, one-million trillion kWh electricity. You may wonder: how is such huge energy generated? (Fig. 7).

To answer this question, we need to first of all understand how the sun was born. Before the sun came about, there were clouds of hydrogen compressed to rotate due to gravitation. The gravitational force accelerated the motion of atoms. This resulted in rising temperature of the hydrogen cloud (the gravitational energy was converted to kinetic energy and then to heat because of mutual friction). During the initial rotation process, a small and relatively harmonious ball was formed. As more hydrogen was absorbed, the small ball grew bigger. In this state, the hydrogen around the surface area all tended to rush in, yet the hydrogen active at the core couldn't get out. As a result, the core temperature went up constantly and so did the density (Fig. 8).

Fig. 6 A view of the sun from the earth

Knowledge Link

Hydrogen

Hydrogen is the first element in the periodic table of elements. Abbreviated as H, it is the lightest chemical element. Hydrogen is widely distributed in nature in that it accounts for about 11% of the mass of water which functions as its "warehouse"; soil contains about 1.5% of hydrogen; oil, natural gas, animals and plants all contain hydrogen. Yet there is not much hydrogen in the air—only about 1/200,000 of the total. Hydrogen is the most common element in the universe, accounting for about 75% of its mass (Fig. 9).

When temperature of the core exceeds 10 million degrees Celsius, the hydrogen at the core will be squeezed to the degree that the nucleus sticks together. The stickiness is magical as two atomic nuclei induce fusion reaction which produces a helium nucleus and a neutron. The key is that the products are lighter than their parent nuclei. This means that the hydrogen only lost a small amount of mass. Then where is the lost mass?

As we know, energy and mass are mutually convertible. The mass lost is released and converted to energy. This conversion is remarkable, because the conversion coefficient of mass and energy is the square of the speed of light ($E = mc^2$, energy =

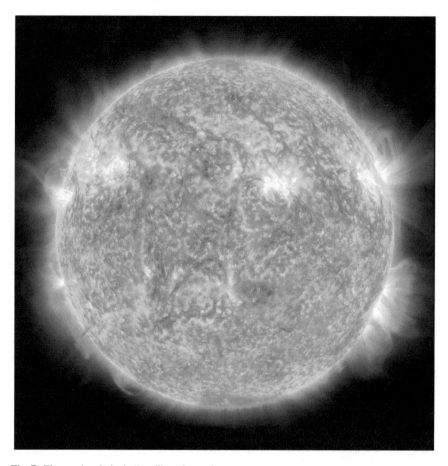

Fig. 7 The sun is relatively "cool" on the surface

mass × speed of light). Therefore, this very little amount of mass is able to generate huge energy.

The energy generated produces an outward force that reaches a balance with the inward force produced by the hydrogen that tends to squeeze in.

The sun has existed for around 5 billion years since this birth. It will continue to give out light for 5 billion years. The sun transmits light and heat continuously to the earth. Plants on the earth absorb light and carry out photosynthesis to form organic matter with water and carbon dioxide and release oxygen at the same time. The process offers abundant food and oxygen for human beings and animals.

Fig. 8 Universal gravitation attracts hydrogen to rotate towards the core

Knowledge Link

Photosynthesis of Plants

Photosynthesis is the process that green plants containing the chloroplast use photosynthetic pigments under visible light to transforms carbon dioxide and water into organic matter and generates oxygen as a byproduct. It is also the process that converts luminous energy into chemical energy to fuel organisms' activities. Photosynthesis is the result of a series of complex metabolic reactions, the foundation for survival of the living beings, and a pivotal medium for the carbon–oxygen balance (i.e. the balance between carbon dioxide and oxygen) on the earth (Fig. 10).

A large quantity of animals and plants were buried deep due to the Tectonic movement. After a quite long period of time, they form coal, oil and natural gas (all called the fossil fuel). Therefore, the fossil fuel which we consume for a large

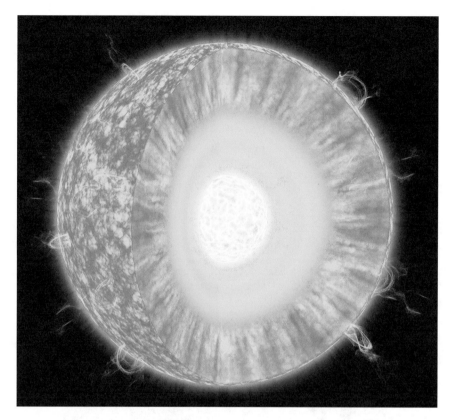

Fig. 9 The core of the sun has the highest temperature

amount originally comes from the sun. Now let's have a detailed understanding of the "Three Musketeers"—coal, oil and natural gas (Fig. 11).

Knowledge Link

Most of the Energy Resources Come from the Sun

The sun is the largest celestial body in the Solar System. It has a diameter 109 times that of the earth and its mass is 330,000 times that of the latter. Actually, the sun possesses 99.8% of all the mass in the Solar System. In addition to offering energy to the earth, the sun is also essential for the generation of wind energy, water energy, bioenergy and mineral energy etc. Fossil fuels, coal, oil and natural gas etc. are actually the solar energy that derived from ancient living beings. Biomass is the chemical energy stored inside the plant and converted from solar energy through photosynthesis. Besides, hydroenergy, wind energy, tidal energy and ocean energy are all converted from solar energy. It is fair to

Fig. 10 The plant is carrying out photosynthesis

say that the vast majority of the energy humanity need comes from the sun directly or indirectly (Fig. 12).

3 "Three Musketeers" of Fossil Fuels

Coal, oil and natural gas are all minerals formed in the long evolution process of ancient animals and plants. They are available for large-quantity exploitation and convenient for transportation and storage. In addition, they have high energy density

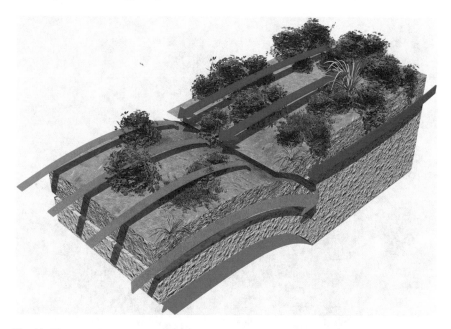

Fig. 11 The tectonic movement buries deep the plant inside earth

and are safe for use. Since eighteenth century, these three forms of energy have been the "Three Pillars" of the modern civilization.

Large-scale exploitation of coal began during the industrial revolution. Since then, it had been ranking top in global energy consumption till 1966 when coal output plummeted and dropped to the second to oil. In 2000, coal output stood at 3.1 billion tons, accounting for 25% of global energy consumption (Fig. 13).

Knowledge Link
Coal

Coal is regarded as the "black gold" and the "necessity of industry". It has been one of the major energy options for humanity since the eighteenth century. In the twenty-first century, however, the value of coal plummeted compared to what it was in the past. Yet judging from the current situation, coal will still remain one of the indispensable energy options for humanity in production and life activities. Coal supply is closely associated with stable development of a country's industry and even various sectors of society. Thus coal supply security is a critical aspect of a country's energy security (Fig. 14).

Oil has the advantage of large heat capacity per unit mass, full burning, convenient transportation, light air pollution etc. It is a sticky fluid in deep brown, crowned as "industry blood". Large-scale exploitation of oil began in 1885, and the output

Fig. 12 The vast majority of the energy humanity need comes from the sun directly or indirectly

reached 3.55 billion tons of standard oil in 2000, accounting for 40% of the world total (Fig. 15).

Knowledge Link

Oil

Oil in Chinese is pronounced as "Shiyou", a term created by Shen Kuo, a scientist in Beisong Dynasty. Oil is mainly composed of a mixture of various alkanes, cycloalkanes, aromatics in the upper crust of the earth. It is mainly

Fig. 13 Coal is regarded as the "black gold"

used as fuel and gasoline as well as raw material for chemical industrial products, such as solvent, chemical fiber, insecticide and plastic etc. The ancient Egyptians and the Babylonians exploited and used oil very early (Fig. 16).

Although found very early, natural gas is hard to transport, and requires large investment and long duration of ROI. The natural gas industry mostly lags behind the oil industries in many countries. In recent year, natural gas has gained the advantages of high heat value and low pollution along with technological advances in liquification, storage and transportation. The market enjoys bright prospects. According to estimates, the future natural gas heat calculation will exceed oil output, take a same 30% market share as oil in energy mix and gradually replace oil (Fig. 17).

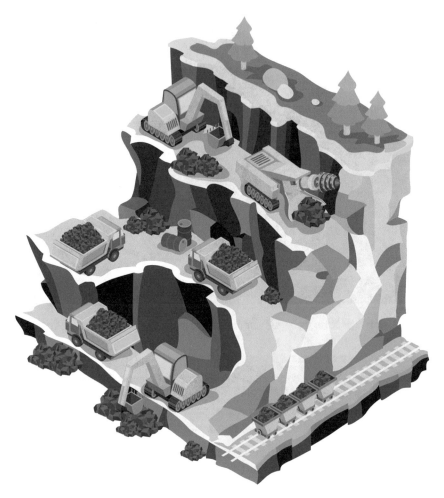

Fig. 14 Coal mining

Fig. 15 Oil is praised as the "blood of industry"

air. These advantages bring safety of the fuel as it will diffuse upward instead of forming explosive gas (Figs. 18 and 19).

Currently, coal, oil and natural gas altogether account for over 85% of the world energy consumption. Just a novel put it, they are the "Three Musketeers" in the Kingdom of Energy. However, like any hero, they also have an Achilles' heel, i.e. limited deposit of the raw material. As humanity's demand on energy is growing, fossil fuels are gradually running out. The world is facing severe energy crisis.

According to statistics of the World Energy Conference, the current coal reserves of the earth are estimated to be available for 200 years, oil for 40 years and natural gas for 60 years. This means that the three are available for human consumption for only around 200 years (Fig. 20).

Due to allocation, we can serve no more gasoline today (图说明文字).

To address energy crisis, humanity urgently need to develop new energy to fill the gap left by fossil fuel energy shortage and ensure long-term sustainable energy supply for humanity.

Fig. 16 Oil extraction

4   The Kingdom Awaits Successors

As the fossil fuel is gradually depleted, humanity have intensified exploitation of solar energy, wind energy, hydro energy, ocean energy, tidal energy, bio-mass and geo-thermal energy etc.

Solar energy is widely utilized. It falls into two categories: photovoltaic and solar thermal. Photovoltaic energy directly transforms energy from the sun into electricity; the solar thermal heats water to hot water or steam by absorbing energy from the sun. The hot water or steam is available for direct use, just like that from solar water heater at home. Meanwhile, it can push the steam turbine to generate electricity. Anyway, regardless of the category, the ultimate goal remains the same, i.e. collecting the energy offered by the sun to the earth for our daily use (Figs. 21 and 22).

The left is a solar panel with the photovoltaic system; the right is a solar water heater with the solar thermal system (图说明文字).

Wind energy is a kind of clean and renewable energy. Humanity has been used wind energy since ancient times before Christ. Wind energy essentially belongs

Fig. 17 Flames of natural gas combustion

to solar energy. When the sun shines on the earth, temperature difference in the surface occurs due to difference in heat. The difference in temperature brings about atmospheric convection which results in wind. According to statistics, only 2% of the solar energy which reaches the earth can be converted to wind energy. Yet the overall amount is remarkable—global wind energy is 10 folds of water energy in terms of total volume (Fig. 23).

Water energy refers to river and marine energy. The water energy here refers to river energy. The strength of water energy mainly depends on the landform and physical features of the river. China has vast land expanses and a large number of rivers, topographically high in the west and low in the east with relatively big difference in elevation. These make China's water energy resources rank first in the world. Marine energy produces tidal energy, wave energy, marine current energy, ocean thermal energy etc. Among them, tidal energy and wave energy are commonly exploited as a result of difficulty in exploiting other forms (Fig. 24).

Biomass energy is produced on the basis of biomass. It can store solar energy in biomass as chemical energy (photosynthesis). Biomass energy can be directly utilized as fuel. Drilling wood to make fire and cooking wood to make charcoal back to ancient times both demonstrate the use of biomass. Today, humanity pays high attention to biomass and has produced new forms of biomass as liquid or gaseous fuel, including bio-diesel, bioethanol, biochar and others (Fig. 25).

Fig. 18 Natural gas extraction

Knowledge Link

Biomass

All organic matter that can grow is biomass, including plants, animals and microorganisms and their waste. Representative biomass includes agricultural plants and their waste, wood and its waste and animal waste. During production activities in agriculture and forestry, biomass mainly refers to lignocellulose that excludes grain and fruits, such as straw and trees, remnants from farm produce processing, the waste and residues from agriculture and forestry, animal droppings and waste from husbandry operations etc. (Fig. 26).

Fig. 19 Fossil fuel is running out

Fig. 20 Sign of no more gasoline supply

Fig. 21 Solar energy

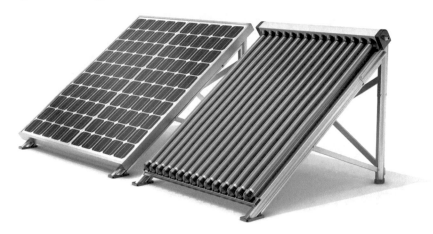

Fig. 22 Two ways to use solar energy

Fig. 23 Wind energy

Fig. 24 Water energy

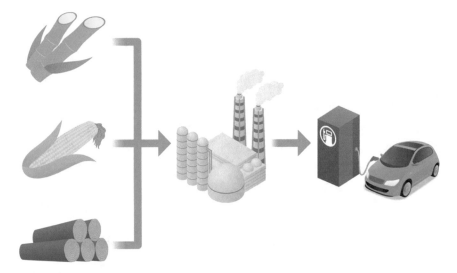

Fig. 25 Biomass energy

Geothermal energy is the heat reserved in the earth. The temperature at the center of the earth reaches 5000–8000 °C. Eruptive volcanos and hot springs show us the huge heat contained within the planet. Humanity utilizes hot water and steam from the underground to generate electricity. The energy has not been extensively developed due to high costs in geothermal exploitation and the worry that geothermal exploitation will cause sinking of the ground (Figs. 27 and 28).

5 The King in the Future

Solar energy, wind energy, hydroenergy, marine energy, tidal energy, biomass energy, geothermal energy etc. tend to be the new "Musketeers". However, they still can't replace the fossil fuel due to more weather or geographical constraints. In addition, the power generated by these resources is insufficient to meet daily demand of humanity. Therefore, humanity urgently needs to find new energy comparable to fossil fuels. As a result, nuclear energy comes to the limelight.

Atomic bomb and hydrogen bomb represent two forms of nuclear energy release. However, humanity also worry about the consequence caused by nuclear energy. If all nuclear weapons of all countries are used to attack the earth, humanity will surely suffer tremendous catastrophe. This is something we never hope for. In order to prevent the nuclear war and extinction of species by various nuclear weapons, 59 countries including the UK, US and the Soviet Union signed *The Treaty on the Non-Proliferation of Nuclear Weapons* which aims to prevent nuclear proliferation

Fig. 26 Biomass is a kind of green energy

and enhance international cooperation on peaceful use of nuclear energy (Figs. 29 and 30).

Humanity has imposed control on nuclear weapons, yet peaceful use of the nuclear energy is still proceeding. Research analysis shows that the reason for huge damage from the atomic bomb and the hydrogen bomb is that their reaction is uncontrollable and that all energy is completely released immediately. If we can control the reaction process to slow down the release, the nuclear energy will be usable for daily purposes.

Scientists therefore start to study the controlled nuclear reaction. Thanks to the study, humanity has mastered the reaction principles for nuclear fission (Figs. 31 and 32).

Despite huge energy produced by nuclear fission, life cycle of the energy is not long enough (materials for the fission are usable for 1000 years). Therefore, nuclear fusion still can't be the major energy resource. The mission then goes to nuclear fusion. Yet the determinant factor for the future energy king still depends on whether or not the nuclear fusion energy can last long enough.

Fig. 27 The earth's center is like a firing ball

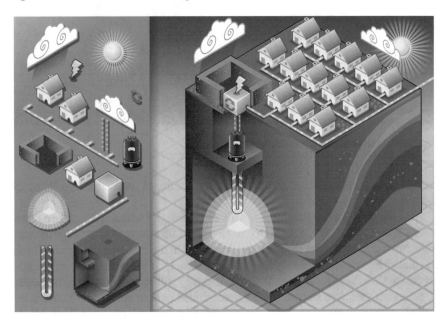

Fig. 28 Geothermal utilization

Fig. 29 Mushroom cloud after the detonation of an atomic bomb

Nuclear fusion materials can be extracted from seawater. Seawater resources on the earth are abundant. According to estimation from experts, fusion energy can last for 10 billion years. Now the earth is aged around 4.6 billion years old. Compared with life expectancy of the earth, nuclear fusion energy is almost inexhaustible (Fig. 33).

With controlled nuclear fusion power generation, humanity will be able to address energy problems in a fundamental way. Nevertheless, in face of huge difficulty, no country in the world has been managed to a build nuclear fusion power plant.

Currently, humanity will continue to rely on fossil fuels until they are depleted one day. Of course, humanity are also improving efficiency of various new energy resources to reduce reliance on fossil fuels. For the future after a thousand years, humanity can only depend on nuclear fusion energy for survival. No matter how difficult it is, controlled nuclear fusion must be a success (Fig. 34).

Fig. 30 Mushroom cloud after the detonation of a hydrogen bombing

Fig. 31 Nuclear power plant (nuclear fission)

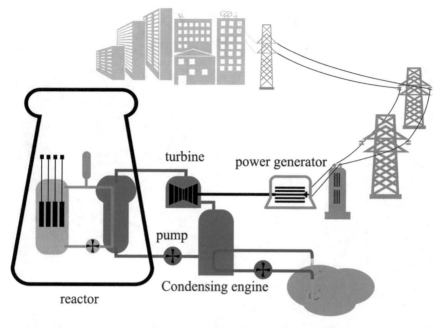

Fig. 32 Power generation process of the nuclear fission power plant

Fig. 33 The earth is endowed with rich seawater resources

Fig. 34 Nuclear fusion energy is the future of humanity

All About the Nuclear Status Quo

Suzhen Zhang

Abstract As a latecomer of energy resources, nuclear energy has been impressive ever since it came into being. You must want to know, out of curiosity, why nuclear energy is so powerful. In this chapter, we will present the answer in details to you.

1 A Big World in a Small Nucleus

Unlike the "predecessors" such as solar energy, wind energy and hydroenergy, nuclear energy was discovered by humanity very late. Initially, human beings knew nothing about the structure of matter. As the knowledge about the world goes deep, nuclear energy, which exists in the micro world, gradually come to the spotlight. A review of its history shows that nuclear energy dates back to 1897 when a British physicist (Thomson) discovered electron. Before the discovery, people generally believed that the world was made of indivisible particles of matter. The particle is called atom, a word meaning indivisible in Greek. People believed that the atom is like an indivisible solid glass marble without any internal structure.

> **Knowledge Link**
> **Dalton's Atomic Theory**
>   The assumption that "matter is made of atoms" was put forward long ago. Yet it was not until the eighteenth century, especially from the late eighteenth century to the mid nineteenth century, that people came to know more about the atom thanks to the mid nineteenth century when modern industries and science were booming, science and production activities were developing rapidly, and massive chemical and physics experiments were advanced (Fig. 1).

S. Zhang (✉)
Institute of Plasma Physics, Chinese Academy of Sciences, Hefei, Anhui, China
e-mail: yihco0101@sina.com

© Zhejiang Education Publishing House 2021
B. Wan (ed.), *Man-Made Sun*, China's Big Science Facilities,
https://doi.org/10.1007/978-981-16-3887-9_2

Fig. 1 John Dalton
(1766–1844)

The success that the Atomic Theory develops into a scientific concept is attributed to a British scientist named Dalton. In 1803, Dalton proposed the "Atomic Theory" in which he believes the smallest unit of the world is atom; atom is unitary, independent, indivisible and stable amid chemical changes and atoms of the same kind share same qualities.

Dalton's "Atomic Theory" explained many basic chemical laws and was thus quickly accepted by people at that time (Fig. 2).

After the discovery of the electron, people came to realize that the atom is divisible and that the atom has internal structure. As the electron carries the negative charge, there must be the particle that carries the positive charge. Only in this way, the charge balance can be maintained and the atom remains electrically neutral. However, people soon raised new questions: apart from electrons, what else does the atom contain? How does the electron work inside the atom? What are the particles in the atom that carry the positive charge? How is the positive charge distributed?

How does the electron that carries the positive charge interact with the particle that carries the negative charge? This long list of questions went to physicists. They then proposed various atomic models in full imagination according to scientific practice and experimental results (Fig. 3).

Fig. 2 "Atomic Theory" holds that the atom is comparable to a solid glass marble

In 1904, Thompson proposed the model of "Dried-grape Cake". He believes that electrons are evenly distributed in spheres with positive charges, just like dried grapes dotted in a piece of cake. This model explains not only why atoms are electrically neutral, but also the cathode ray and the phenomenon that metals emit electrons when exposed to ultraviolet radiation. What's most remarkable is that this model enables estimation of the atomic size. The model was hence widely accepted by physicists at that time (Fig. 4).

In 1895, Rutherford came to the Cavendish Laboratory in the UK and became Thomson's graduate student. In 1910, Rutherford confirmed the existence of the nucleus through experiments. In 1911, Rutherford established a nuclear atomic model. The atomic model proposed by Rutherford is like a solar system. Nuclei with positive charges are like the sun and electrons with negative charges are like the planet orbiting the sun.

In this "solar system", the force that controls the "planets" is from the electromagnetic interaction. Rutherford believed that atoms consist of nuclei and electrons. The nucleus carries the positive charge and the electron carries the negative charge.

Fig. 3 The atomic model
proposed by Thompson

Structure of the Atom

Thomson's Model of the Atom

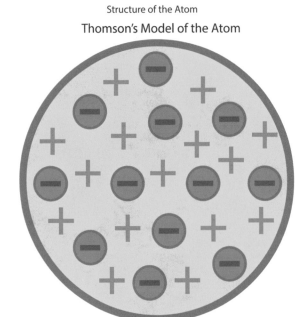

Fig. 4 Thompson's atomic model can be compared to the "dried-grape cake"

Fig. 5 Ernest Rutherford (1871–1937)

Their charge volume is equal. Most of an atom's mass is in its nucleus. Although massive, the nucleus is very small. If the atom is compared to a football field, the nucleus will be like a drop of dew on the field (Fig. 5).

Rutherford's theory attracted attention of a young Danish man who named Bohr. In 1912, Bohr studied in Rutherford's laboratory where he researched his atomic theory. Bohr applied Planck's quantum hypothesis to the energy inside the atom. Based on Rutherford's model, he proposed the quantized orbit model to address stability issue of the atomic structure.

Later on, Rutherford further discovered that the nucleus consists of smaller particles. He proved the existence of protons through experiments and predicted the existence of neutrons. Then, his student Chadwick proved the existence of neutrons through experiments. In same the year when Chadwick discovered the neutron, Heisenberg, a German physicist, proposed a model of nucleus structure based on protons and neutrons (abbreviated to the proton-neutron model).

2   Strong Energy Inside a Small Body

The proton-neutron model explains well the relationship between relative atomic mass and atomic number, thus quickly accepted by scientists. However, they were still puzzled by one question—how do protons and neutrons form the nucleus?

In the nucleus, all protons and neutrons are in a narrow space measured only 10–15 m in diameter. Protons carry positive charges with strong repulsive force. In this situation, the nucleus would be extremely unstable and spread all over due to the electrostatic repulsion. Yet the reality is the opposite, instead of scattering all around, nuclei are firmly connected (Figs. 6, 7, 8 and 9).

At that time, people only knew universal gravitation and electromagnetic force. Yet this reality failed to be explained by the gravitation in that it is impossible for such small particles to produce such strong binding force. Electromagnetic force would only lead to dispersion of particles. In order to explain this phenomenon, physicists began to hypothesize that a force stronger than the static one exists in the nucleus.

Fig. 6 The "solar system" model proposed by Rutherford

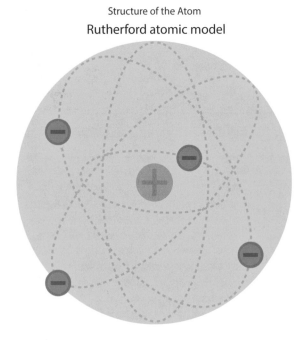

Structure of the Atom
Rutherford atomic model

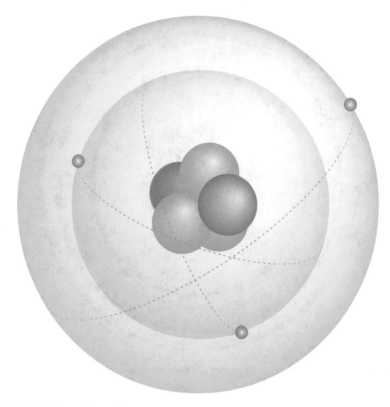

Fig. 7 Bohr's electronic quantized orbit model

Knowledge Link
Universal Gravitation

Universal gravitation refers to the interactive force between any two physical bodies. It was discovered by Newton in 1687 and defined as the law of universal gravitation. The law states that there exists an attractive gravitational force between any two bodies; the force is directly proportional to the product of their masses and inversely proportional to the square of their distance.

People define this force as nuclear force within the range of the nucleus, which is much stronger than Coulomb force. In addition, nuclear force is irrelevant to the charge, i.e. either charged electrons or uncharged neutrons are all affected by nuclear force. Nuclear force is the third force known by human beings up till now. It perfectly

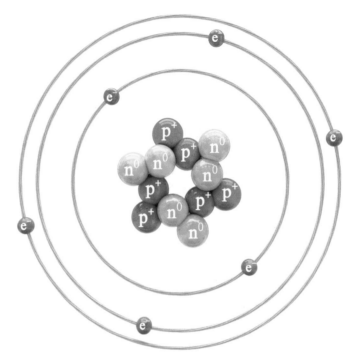

Fig. 8 Proton-neutron model (p is for proton; n is for neutron; e is for electron)

explains how protons and neutrons are firmly constrained to form stable nuclei. Later experiments also proved this guess that nuclear force indeed exists (Fig. 10).

Knowledge Link

Characteristics of the Nuclear Force

Nuclear force exists within the nucleus. It is much stronger than the Coulomb force, around 100 times of the later. Nuclear force is irrelevant to the charge. The nuclear force among protons and neutrons, and between protons and neutrons, is equal. The nucleus can't be approached infinitely. This means that nuclear force is manifested in the form of repulsion in addition to gravitation. Generally speaking, the nucleus with the distance between each nucleus that measures from 0.8 to 1.5 fm (1 femtometer equals 10^{-15} m) is manifest as gravitation, and the nuclear force with the distance less than 0.8 fm is manifest as repulsion. This is why the nucleus is not bound together. When the distance exceeds 1.5 fm, the strength of nucleus force will plummet. Each nucleus, therefore, only affect its neighboring nucleus with nuclear force.

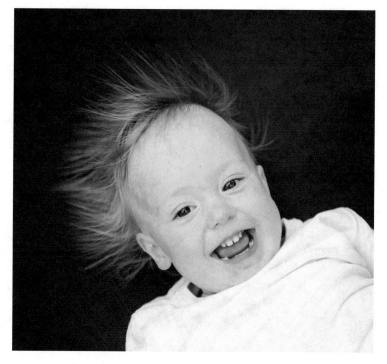

Fig. 9 Electrostatic repulsion makes your hair stand on end

Up to this stage, scientists have obtained clear understanding of the nucleus. They have clearly known that there exists huge energy inside the nucleus—because the force that binds nuclei together is very powerful, the relevant energy must be also powerful. This lays a solid foundation for nuclear energy exploration.

3 The Nuclear Box Is Opened

Under the background that the science community has not yet fully understood the nuclear force process, when they determined the nucleus mass, scientists found that there was huge difference before and after the nucleus binding. For example, the atomic mass of helium is measured 4.002602 amu; the core of helium is made of four nuclei (two protons and two neutrons) and yet the total mass of the four nuclei is 4.032980 amu. Very strangely, there is "loss" of mass before and after the nucleus binding process; the lost amount is 0.030378 amu (Fig. 11).

Fig. 10 Nuclear force bonds together protons and neutrons

Knowledge Link

Atomic Mass Unit

Atomic Mass Unit is adopted across the globe as a standard to calculate the mass of micro molecules. It is defined as 1/12 of the mass of an atom of carbon-12 in the symbol of u. 1 u = (1.6605402 ± 0.0000010) × 10^{-27} kg is based on experimental results. In writing of the relative atomic mass, the unit will be omitted with the atomic mass unit considered as the default unit (Fig. 12).

This means the mass of a single nucleus is bigger than the mass of the nuclei bound together. The nucleus is forced to bind firmly together, during which process the mass is reduced. This phenomenon is called "mass loss". According to the mass

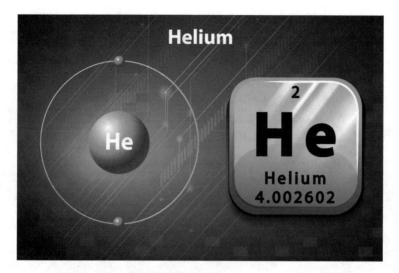

Fig. 11 The atomic mass of helium is 4.002602 amu

Fig. 12 Atomic mass unit is defined as 1/12 of the mass of an atom of Carbon-12

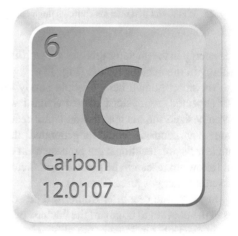

energy equation, when the energy of an object changes, its mass will also change accordingly. Light speed is a large parameter (about 3×10^8 m/s), this implies that a small amount of mass can be transformed into huge energy.

Humanity thus realized that the nucleus actually contains tremendous energy. This undoubtedly provides a strong tool to change the world, and the significance is comparable to the invention of fire. However, despite the discovery of the strong "energy box", humanity didn't find the way to open the box. Some Scientists even believed that humanity would never open this box. For example, Rutherford who found atomic nuclei still believed before he passed away in 1937 that more energy must be consumed to obtain the particles that bombard the nucleus in order to release

Fig. 13 It is found that nucleus contains huge energy

the energy in the nucleus; this was not worth the candle; humanity would never exploit energy in the nucleus for use (Fig. 13).

Yet smart human beings would surely think of ways to open the box in hand. As expected, a physicist named Szilard was inspired from a science fiction and hypothesized the nuclear fission chain reaction: during the nuclear fission, two or three new neutrons would be generated; the new neutron would bring about more nuclear fission reactions; the process went on and on, interconnected like the chain. In follow-up research, he precisely predicted the result of the chain reaction and the potential energy source (Fig. 14).

In 1938, one year after Rutherford passed away, a German scientist named Hahn used the neutron to bombard the nucleus of uranium and achieved the nuclear fission chain reaction. On January 6th 1939, Hahn published a series of experiment results and immediately caused sensation in the physics community. Scientists realized that the nuclear fission chain reaction opened the box of nuclear energy which would release huge amount of energy (Fig. 15).

During the nuclear fission reaction, the nucleus of uranium lost some mass which would release huge amount of energy. Theoretically, the explosive force produced by Uranium-235 in continuous fission reactions is 20 million times that of TNT explosives of the same mass. The fission interval is less one millionth of a second. That is to say, huge amount of energy is released at an extremely short time. This means the fission could turn into an astonishingly powerful bomb.

Fig. 14 Chain reaction is activated like the domino effect

Knowledge Link
TNT Explosive
  TNT is a strong explosive. Pure TNT is colorless needle-like crystal. Industrial TNT comes in the form of yellow power or scale. Poorly soluble in water, TNT explosive can be used for underwater blasting with a detonator. Because of its strong power, it is frequently used as a secondary explosive. After the explosion, the oxygen balance is negative and toxic gases are produced. TNT explosive is essentially stable and hard to explode and will not detonate even if hit by a bullet.

4   The Astonishing Atomic Bomb

In 1939, within several weeks after Hahn published his research results in nuclear fission reactions of the uranium nucleus, scientists in many countries proved this discovery through experiments and further proposed conditions for slow process of

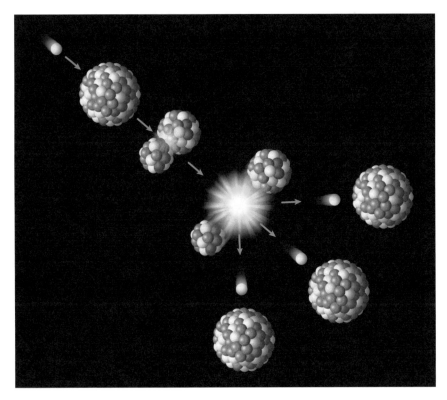

Fig. 15 Chain reaction of nuclear fission

such fission reaction which presented broad prospects for exploiting this new energy resource to generate wealth for humanity. Nonetheless, like many other new scientific discoveries, nuclear energy was initially exploited for military use, i.e. making the extremely powerful atomic bomb (Fig. 16).

Knowledge Link

History of Atomic Bomb

In April 1939, Nazi Germany brought six atomic physicists to Berlin to hold a secret meeting in which the country decided to develop facilities for controlling and exploiting uranium. In summer of the same year, Germany prohibited uranium ore export from Czech under the occupation of Germany and severely banned uranium news reports. In September, Germany gathered together scientists to form a uranium society to launch a secret development project code-named "U Plan". This signifies the beginning of Nazi Germany to develop nuclear weapons.

Fig. 16 Nuclear energy was initially used to make the extremely powerful atomic bomb

The Danish physicist Bohr heard from two escaped German physicists about the precise information that Germany had already started to develop the atomic bomb, and immediately told this to Fermi and other scientists in the US.

This information deeply worried the scientists who knew the huge power of nuclear energy. They understood that humanity would face unprecedented catastrophe if Germany succeeded in making the first atomic bomb. The only way out was to develop an atomic bomb ahead of Germany.

Therefore, Fermi and other scientists, out of historical responsibility and in active coordination, persuaded scientists not to publish research results in nuclear fission and chain reaction to avoid being used by German researchers. Meanwhile, they invited the then scientific icon Einstein to write to President Rosovolt of the US about the huge potential of the nuclear chain reaction and the possibility of making an atomic bomb and to strongly suggest the US to prevent Germany from making the first nuclear weapon. This letter directly gave rise to the significant "Manhattan Project". This three-year "Manhattan Project" brought together top scientists in developed countries at that time with an investment of $2 billion. On 16th July, 1945, the US successfully conducted the first nuclear explosion in the world (Fig. 17).

The atomic bomb is one of the nuclear weapons. It makes use of the photothermal radiation, shock waves and induced radioactivity to cause damage, destruction and large area of radioactive pollution. It is the weapon of massive destruction which is aimed at preventing the other party from realizing strategic goals of military operations. The first atomic bomb in the world was developed by the US. In July 1943, the US set up the Institute of Atomic Bomb. On July 16th 1945, the explosion test of the first atomic bomb was conducted in the Alamoque Desert of New Mexico. Soon after that, the US developed two more atomic bombs, code-named "Little Boy"

9.0 SEC.
N
⊢——⊣ IOO METERS

Fig. 17 The scene in nine seconds after the first atomic bomb denotation

and "Fat Man" respectively. The US carried the two atomic bombs in a bomber and dropped them in Hiroshima and Nagasaki of Japan. On August 15th, Japan surrendered unconditionally, which brought an end to the Second World War (Figs. 18 and 19).

The strong light wave caused by the atomic bomb explosion could blind tens of thousands of people; the high temperature of thousands of degrees Celsius could turn anything into ashes; the radioactive rain could cause death of some people within 20 years after the explosion; the raging wind induced by the shock wave could smash all the buildings into pieces. In the long history of humanity, the discovery and use of hydropower, coal, oil and natural gas were all accompanied by celebration cheers. Nuclear energy is the only one that came to the earth in the image of a catastrophic destroyer. "Nuclear energy" had casted a shadow on human beings since it came out (Fig. 20).

After the Second World War, scientists who participated in the atomic bomb research felt regretful. They started to reflect and engage in the anti-war movement to constantly warn people about the danger of the nuclear weapon. In recent year, as people have become more aware of the danger from nuclear proliferation, they require complete elimination of nuclear weapons. In particular, the voice against various types of nuclear tests has grown increasingly loud. International organizations such

Fig. 18 The atomic bomb code-named "Little Boy"

Fig. 19 The atomic bomb code-named "Fat Man"

Fig. 20 Nagasaki in Japan after the atomic bomb was detonated

as the UN have adopted a series of treaties and agreements to constrain the production and use of the nuclear weapons by countries across the world (Fig. 21).

5 The Benefit of the Nuclear Power Plant

While the US conducted nuclear tests, the Soviet Union also carried out nuclear energy research. In 1943, the Soviet Union decided to develop nuclear weapons and conducted the first nuclear test in August 1949. After the Second World War, countries around the world started to exploit nuclear energy for peaceful purposes, i.e. building nuclear fission stations. This time, the Soviet Union run in front of the US and successfully built the first nuclear power plant in the world in June 1954. The first nuclear power plant of the US started operation in 1957. After that, many countries launched the project to build the nuclear power plant. There have been over 100 nuclear power plants across the world up till today (Fig. 22).

Currently, nuclear power plants mostly use the nuclear fission energy which has considerable advantages over the fossil fuel.

Fig. 21 The UN constrains nuclear weapons through treaties and agreements

Fig. 22 Countries around the world are building nuclear power plants

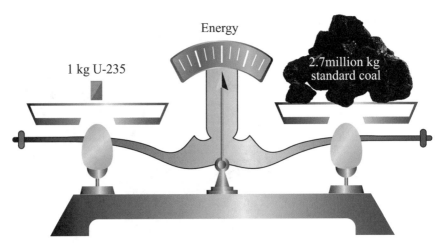

Fig. 23 1 kg U-235 fission releases energy equal to that generated from combustion of 2.7 million kg standard coal

First of all, the energy released in nuclear fission is huge. Fission of 1 kg of U-235 releases energy equal to that generated from the combustion of 270 kg standard coal (Fig. 23).

Second, nuclear fission energy produces little pollution. Combustion of fossil fuels produces carbon dioxide that leads to the greenhouse effect. Thermal power plants where electricity is generated by coal also keep emitting hazardous substances such as carbon dioxide and nitric oxides to the air. Meanwhile, a small amount of radioactive substances such as uranium, titanium and radium contained in the coal also fall along with the smog in surrounding areas and cause pollution. Nuclear power plants, built with multiple layers of protection, emit few polluting substance and much less radioactive pollutants than thermal power plants. According to statistics, a nuclear power plant in normal operation exerts less radioactive effect on residents in surrounding areas than one time of X-ray (Fig. 24).

Because of numerous advantages of the nuclear power, there have been continuous call and actions for developing nuclear power around the globe. Countries in the world initially remained extremely passionate about nuclear power stations. Especially from 1960 to 1970s, power demand had doubled as a result of rapid industrial development. During the period, the US, the Soviet Union the UK and the France etc. made ambitious nuclear power development plans. Nuclear power units of the US amounted to over 100 just within a decade. Germany, Japan, India and Brazil etc. also purchased whole sets of equipment to build nuclear power plants. In particular, after the first oil crisis in 1973, nuclear power plant development reached the first peak as countries tried to be independent of the Middle East in oil production.

The golden era of nuclear power had lasted for a decade. However, since 1970s, the world has been experiencing continuous slowdown in installed nuclear power generation capacity. Installation plans has been constantly cancelled; over one third of

Fig. 24 Fossil fuel combustion causes environmental pollution

all plans after 1970 have been ultimately cancelled. The main reason is the increasing economic cost, coupled with the falling price of fossil fuels such as oil and natural gas. Nuclear power plants thus became less attractive under this background. Furthermore, large scale of energy conservation measures and industrial restructuring lead to delay or cancelation of many new nuclear power plant development programs (Fig. 25).

Despite all the slowdown, exploitation of nuclear power is still a major trend in the world. Electricity generated by nuclear energy currently accounts for 16% of the global total. Nuclear power plants have made huge contribution to energy use of humanity. As a developing economy, China still has rapidly growing demand for energy to support social and economic development. Meanwhile, as China is facing the pressure to tackle haze, developing nuclear energy will be inevitable for China (Fig. 26).

Fig. 25 Continuous improvement in safety standards of nuclear power leads to increasing economic costs

Ling Ao Nuclear Power Plant   Hong Yan River Nuclear Power Plant   Ning De Nuclear Power Plant

Yang River Nuclear Power Plant first-stage project   Tai Shan Nuclear Power Plant

Fig. 26 Nuclear power plants are developing rapidly in China

The Legend of Artificial Sun

Yaqin Li

Abstract To use nuclear energy is a double-edged sword. It will help address energy crisis for human beings if used as a kind of energy resources, or it will destroy human civilization if used as a weapon.

1 Nuclear Safety Faces Worries

As the shadow of the atomic bomb is yet to disappear, new nuclear panic is coming. While humanity is trying hard to develop controlled nuclear fission energy and build nuclear power plants one after another, horrifying nuclear accidents happen incessantly.

At 4:00 March 28th 1979, the alarm bell suddenly went off at Three Mile Island Nuclear Power Plant in Pennsylvania of the US. The turbine stopped spinning; pressure and temperature of the reactor core suddenly went high; most parts of the reactor were burned down and some radioactive substances were released as a result. Temperature of the core went down till six days later. Although an explosion was avoided, the reactor had been paralyzed. The accident didn't cause damage and casualty to the environment and residents in the surrounding community; no obvious radioactive effects were discovered. Yet it left an indelible shadow. Three Mile Island Nuclear Power Plant accident was the severest nuclear accident that shocked the entire country. After the accident, about 200,000 residents were evacuated from the area (Fig. 1).

The Chernobyl Nuclear Power Plant is the first nuclear power station built by the Soviet Union. Back to the glory days, the Chernobyl Nuclear Power Plant is the pride of the Soviet people. It is believed to be the safeties and the most reliable nuclear power plant in the world. Nonetheless, a huge blast brought all the glories to an end. The No. 4 reactor suddenly caught fire during a half-baked test and caused explosion. The explosion completely destroyed all facilities and leaked out over eight tons of radioactive substances. The radioactive dust drifted with wind to Russia, Belarus and

Y. Li (✉)
Institute of Plasma Physics, Chinese Academy of Sciences, Hefei, Anhui, China
e-mail: yihco0101@sina.com

© Zhejiang Education Publishing House 2021
B. Wan (ed.), *Man-Made Sun*, China's Big Science Facilities,
https://doi.org/10.1007/978-981-16-3887-9_3

Fig. 1 Three Mile Island Nuclear Power Plant in the US

Ukraine which were polluted by nuclear radiation. The area polluted by radioactive dust totaled 200,000 m², making it the most serious nuclear accident up till today (Fig. 2).

The reason for those two nuclear accidents, according to investigations, is mainly human error during operations. Since then, people started to pay high attention to safety of nuclear fission power plants, improve safety standards, design sufficient protection devices, follow very severe operation standards to effectively prevent nuclear accidents (Fig. 3).

It's easy to avoid "human disaster", yet hard to prevent "natural disaster". On March 11th 2011, Japan was struck by a magnitude 9.0 earthquake along with a fifteen-meter tsunami induced. The tsunami caused blackout of the Fukushima Nuclear Power Plant and led to a serious nuclear accident: a large amount of hydrogen was ejected; the reactor roof was ripped off; nuclear substances spread around; the burning fire melted part of the reactor core. Although no casualty resulted from the nuclear radiation, the psychological damage to residents of the surrounding areas is far greater than the nuclear radiation.

These nuclear safety accidents have severely undermined people's confidence in nuclear energy. Whether the public accept or not and to what degree they accept have become important factors to consider for nuclear development; some countries even halt nuclear power development. To ensure safety of the nuclear power, safety measures of nuclear power plants have been constantly improved. This also leads to prolonged construction period, increasing investment and undermined economic competitiveness.

Fig. 2 The radiation amount near the Chernobyl Nuclear Power Plant is still high

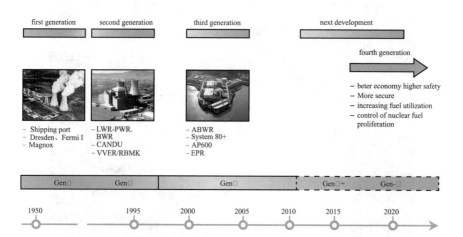

Fig. 3 Human beings continuously advances technologies of nuclear fission reactors

2 King of Nuclear Weapons—Hydrogen Bomb

Before we talk about the use of nuclear fission energy, we need to first of all understand the hydrogen bomb. Hydrogen bomb explosion is an uncontrollable nuclear fission reaction and it is much stronger than the atomic bomb. Generally speaking, an atomic bomb explosion is equivalent to tens of kilotons of TNT, yet a hydrogen bomb explosion can be as powerful as tens of millions of tons of TNT (Fig. 4).

During the Cold War, the US successfully conducted the explosion test of the first hydrogen bomb in the history in March 1954, equivalent to 15 million tons of TNT. To confront he US, the Soviet Union developed the massive "Big Ivan" H-bomb unrivalled in the world. The explosive force was initially planned 100 million tons of TNT equivalent and was reduced to 50 million tons later due to worry that it might impact surrounding cities and residents (Fig. 5).

On October 30th 1961, the 26-ton "Big Ivan" H-bomb explosion test was carried out at the test site of Novaya Zemlya archipelago that covered an area of 82,600 m^2. At 11:32 am, "Big Ivan" was exploded at the test site. The drop plane had flied 250 km before the explosion, yet the huge shock wave travelled faster and pushed the plane up and down. Behind the plane was an unprecedented terrifying mushroom cloud which floated and rocketed spirally 70 km high. This is the most powerful nuclear explosion in the world till today (Fig. 6).

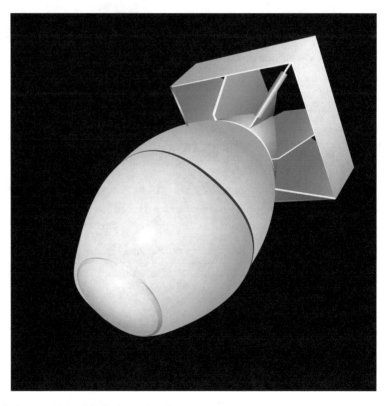

Fig. 4 Configuration of the hydrogen bomb

Fig. 5 Simulation of hydrogen bomb denotation

After the nuclear explosion, all airplanes, missiles, radars and communications equipment within 4000 km were affected to different degrees. The entire communications system of the Soviet Union had been lost for over one hour. The US, opponent of the Soviet Union, also suffered from the explosion, the first victims were the neighboring Alaska and Greenland where the electronic system of the North America Air Defense Commanding Center was mostly paralyzed; radar failed to work and communications were lost. After that, countries in the world never conduct any nuclear test as crazy as this.

3 The Future Energy Depends on Fission

As we know, while safety of nuclear fission power plants continues to be strengthened, humanity starts to focus on another nuclear energy—nuclear fusion.

Nuclear fission energy has already surprised humanity. It is powerful, clean and environmental friendly and free from shortcomings of fossil fuels. Nuclear fusion impressed humanity more excitingly.

First of all, nuclear fusion produces stronger energy. For comparison, 1 g nuclear fission fuel releases energy equivalent to that produced by combustion of 1800 kg oil; yet 1 g nuclear fusion fuel produces energy comparable to that produced by

Fig. 6 "Air drop" of hydrogen bomb

combustion of 8000 kg oil. That is to say, with fuel of the same mass, nuclear fusion produces much more energy than nuclear fission (Fig. 7).

Second, nuclear fusion is cleaner and safer. It won't produce any radioactive substances that pollute the environment, thus very clean. At the mean time, controlled nuclear fusion reactions proceed in an atmosphere of rare gases which eliminate explosion risks, thus very safe.

Finally, raw materials of nuclear fusion in the earth are much richer than those of nuclear fission. Main materials for nuclear fission are uranium and thorium, both non-renewable. Raw materials for nuclear fission can only be used by humanity for around 1000 years. Yet raw materials of nuclear fusion currently are mainly hydrogen isotopes (deuterium, tritium). Deuterium can be extracted from seawater; tritium can be produced by neutron and lithium reactions and seawater also contains a large amount of lithium. According to calculation, one liter of seawater contains 0.03 g deuterium; accordingly, seawater on the earth alone contains 45 trillion tons of

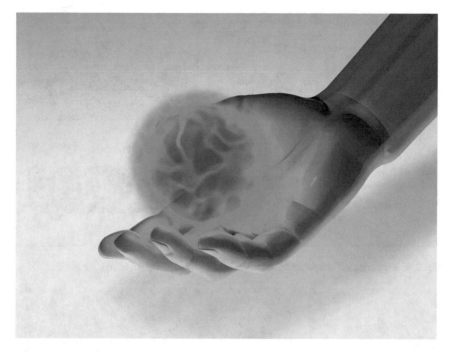

Fig. 7 Nuclear fusion produces more energy than nuclear fission

deuterium. Deuterium in one liter of seawater produces energy after nuclear fusion equivalent to that released by combustion of 300-L oil. Nuclear fusion of the earth is sufficient enough to support humanity for tens of billions of years. Therefore, judging from the current situation, the only reliable energy resource for humanity is nuclear fusion energy (Figs. 8 and 9).

Scientists have researched controlled nuclear fusion technology and tried to build nuclear fusion power plants since very early time. At present, in addition to intensive research, major countries in the world are also enhancing cooperation in this area to advance from nuclear fission to nuclear fusion. This is because humanity will have access to clean, reliable and inexhaustible energy resource if controlled nuclear fusion is realized. It will help protect the ecosystem and environment and at the same time achieve sustainable social and economic development.

4   Daunting Challenges in Nuclear Fusion Research

Humanity has never stopped exploring peaceful use of nuclear energy. In less than a decade since the explosion of the first atomic bomb, peaceful use of nuclear fission energy had already been realized. Operation of the first nuclear power plant in 1954

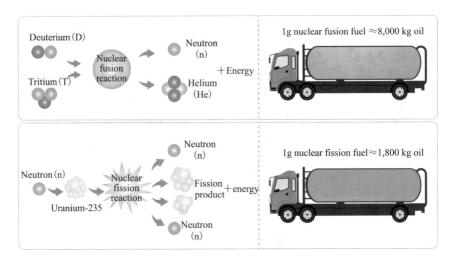

Fig. 8 Energy released in nuclear fusion is the ideal energy for human beings

Fig. 9 Nuclear fusion fuels can be extracted from seawater to generate power

marks the success in achieving human control of nuclear fission reaction (Figs. 10 and 11).

With this experience, humanity estimate optimistically that the day for peaceful use of nuclear fusion energy is coming soon. Scientists started to conduct research on human control of nuclear fusion soon after the hydrogen bomb explosion. They constantly built the nuclear fusion facilities and conducted international cooperation.

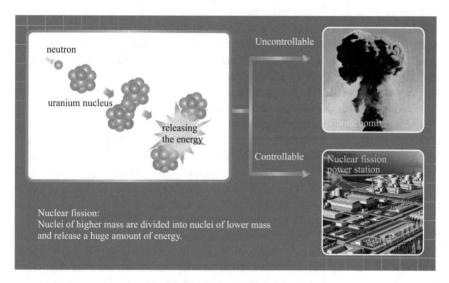

Fig. 10 Peaceful application of nuclear fission is achieved

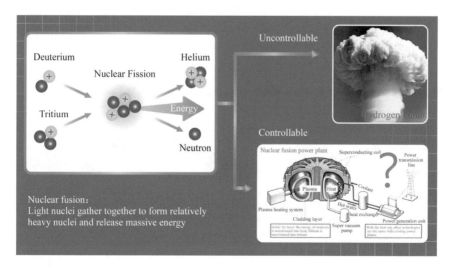

Fig. 11 Peaceful utilization of energy released in nuclear fission has not yet been realized

However, all these efforts encountered failure. People can't help but sigh that nuclear fusion is even harder than climbing the Mt. Evereston bare foot.

If nuclear fission is like breaking a pencil, nuclear fusion will be as hard as recovering the broken pencil to the original with hand. You may puzzle: how is that possible? Right, scientists are turning this mission impossible into something possible (Figs. 12 and 13).

Nuclear fission is hard to achieve due to extremely high requirements for the reaction condition. The sun has a temperature of 15 million degrees Celsius and very high density at the core, under which condition nuclear fusion reactions happen naturally. The hydrogen bomb depends on energy for ignition provided by an atomic bomb installed inside it. If controlled nuclear fusion reaction is to be achieved, the atomic bomb will not be reliable for the ignition in the first place. Secondly, temperature of the earth is not as high as that of the sun; the only option left is a human-made (artificial) environment imitating the sun (Fig. 14).

The current mainstream controlled nuclear fusion research employs the heating approach to provide the initial energy for ignition. According to calculation, if controlled nuclear fusion reaction is to be realized on the earth, the nuclear fusion fuel will need to be heated to 100 million degrees Celsius. This temperature is much higher than that of the sun at the core. People thus give this controlled nuclear fusion project a vivid name—artificial sun (Fig. 15).

During the nuclear fusion research over the past six decades, scientists have proposed new ideas and thoughts, and developed nuclear fusion experimental facilities of various types and kinds. Although it has still not been achieved yet, humanity is always on the way to realize controlled nuclear fusion. We are confident that it will come true in the near future with relentless efforts of the scientist (Fig. 16).

Fig. 12 Realizing nuclear fission is as difficult as breaking a pencil with bare hands

Fig. 13 Achieving nuclear fusion is as difficult as restoring the broken pencil back to normal

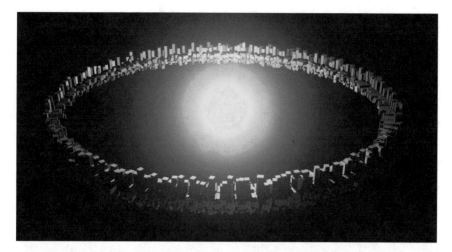

Fig. 14 Controlled nuclear fusion requires an artificial environment imitating that of the sun

Fig. 15 Controlled nuclear fusion project is compared to artificial sun

Fig. 16 Artificial sun is the dream of human beings

How to Create Artificial Sun

Jie Huang

Abstract It is extremely difficult to achieve the artificial sun. Scientists worldwide have been researching on this area for over six decades and still failed to build a nuclear fusion power plant. What are the difficulties? What approaches proposed by scientists to overcome the challenge?

1 The Micro World Presents Truth

To better understand why artificial sun is difficult, we need know something about the micro world first. The reason is that nuclear reaction in the macro world is actually the reaction between individual nucleus in the micro world. Let's fist take a look at the nuclear fission. The material for nuclear fission is usually uranium-235. Uranium is a relatively heavy element that can be found nature and uranium-235 is one of its isotopes. As uranium-235 is relatively heavy, the nucleus contains numerous protons and neutrons. When the uranium-235 nucleus obtains one neutron, distance between each nucleus will increase. Then the mutual gravitation between nuclei becomes smaller than the electronic repulsion between protons. Division of the nucleus easy and ultimately results in nuclear fission reactions (Fig. 1).

The uranium-235 nucleus attracts one neutron and releases two to three neutrons after the fission. The new-born neutron continues to trigger nuclear fission reactions incessantly, making it a nuclear fission chain reaction. This explains why the nuclear fission process is relatively easy (Fig. 2).

J. Huang (✉)
Institute of Plasma Physics, Chinese Academy of Sciences, Hefei, Anhui, China
e-mail: yihco0101@sina.com

© Zhejiang Education Publishing House 2021
B. Wan (ed.), *Man-Made Sun*, China's Big Science Facilities,
https://doi.org/10.1007/978-981-16-3887-9_4

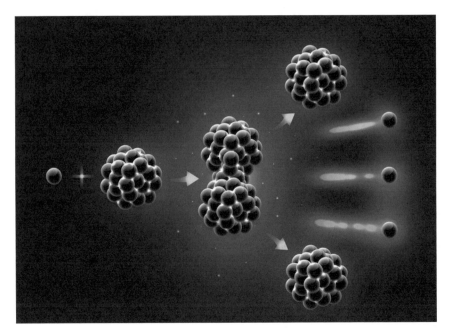

Fig. 1 Neutrons bombard uranium-235 and trigger nuclear fission reactions

Knowledge Link

Uranium-235

Uranium is a chemical element of the period table, atomic number 92. The chemical symbol for uranium is U. The element has three naturally occurring isotopes (uranium-234, uranium-235 and uranium-238), all radioactive. Uranium-235 is the radioactive isotope of 143 neutrons, the only isotope that enables nuclear fission as well as the main fuel used for nuclear reactors and nuclear weapons.

What is the principle for nuclear fusion? Materials for nuclear fusion are generally isotopes of hydrogen—deuterium and tritium. The nuclear fusion reaction is actually the process that the deuterium nucleus combines with the tritium nucleus to form bigger helium nuclei. To make nuclear fusion reaction happen, the first step is to bring the deuterium nucleus and the tritium nucleus close to the extent that nuclear force is effective. When this step is achieved, the huge nuclear force will make the two nuclei combined together to form a heavier nucleus. This is the basic principle for the nuclear fusion reaction (Fig. 3).

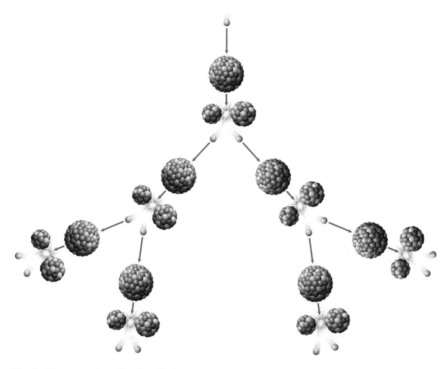

Fig. 2 Chain reaction of nuclear fission

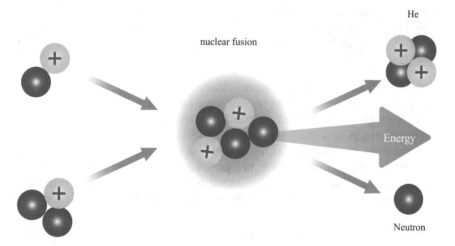

Fig. 3 The principle of the nuclear fusion reaction fueled by deuterium and tritium

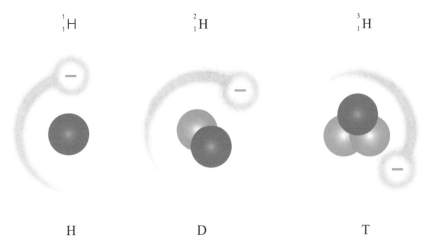

Fig. 4 Structures of protium, deuterium and tritium

Knowledge Link

Protium, deuterium and tritium

   Hydrogen is the chemical element of the period table, atomic number 1. The chemical symbol for hydrogen is H. It has three naturally occurring isotopes (protium, deuterium and tritium). As nuclear fusion reactions happen between deuterium and tritium in most cases, people generally use the two isotopes as nuclear fusion fuels. The deuterium nucleus is made of one proton and one neutron while the tritium nucleus is made of one proton and two neutrons. The later is a half-life (about 11 years) element (Fig. 4).

   The tricky part of the nuclear fusion reaction is that it is very hard to pull the two nuclei close to the point that the nuclear force is effective. The effective distance ranges from 0.8 to 1.5 m. When it's over 1.5 m, the force will plummet.

   The two nuclei both carry positive charges. Two charged particles will display attractive force if same charged, or repulsive force if oppositely charged. There exists electric charge repulsion between particles with the same electrical property and we call it electric repulsion. The electric repulsion is inversely proportional to the distance between particles, i.e. as the distance decreases, the electric repulsion increases.

   In order to make the two nuclei closer enough so that the nuclear force can have an effect, we need to provide them with sufficient energy from the outside. This will also give them enough kinetic energy to overcome the electric repulsion. Then how much energy is needed to make that happen?

2 How to Ignite the Sun?

If we compare the nuclear fusion fuel to firewood, then we will have to first light the wood to make it burn. The problem is that it is hard to light this "firewood". Before the ignition, scientists have to research in what conditions the "firewood" can be lit.

In 1957, the British scientist Lawson proposed an equation after calculation, i.e. Lawson criterion. Lawson criterion sets out a condition for the ignition of nuclear fusion: when product of temperature, density and constraint time of the nuclear fuel exceeds a certain value, nuclear fusion will happen.

Temperature and density are both the concept of the macro world. In the micro world, temperature is actually the motion speed of particles. When temperature rises, the speed also increases accordingly. Density denotes quantity of the particle per unit volume. When density rises, the number of the particle per unit volume also increases. Nuclear fusion in the macro world is to make the nucleus collide and combine in the micro world (Fig. 5).

First of all, we can accelerate the motion speed of nuclei. Higher speed provides relatively strong motion energy so that the nucleus overcomes the electric repulsion

Fig. 5 The nuclear fusion reaction is the process by which two nuclei collide and combine

in between. Increasing speed also means higher temperature in the macro world which offers higher probability for nuclear fusion (Fig. 6).

Second, we can increase the number of the nucleus in a certain fixed volume. In this way, the density of the nucleus will grow, and the probability for collision will be higher. Increasing the number means increasing the density in the macro world, or higher density will increase the probability of the nuclear fusion reaction (Fig. 7).

Finally, if the speed and number are not increased and nuclei are confined in a fixed pace long enough, there will always remain chance for nuclei to collide with each other. This means prolonging the time to confine the nucleus (abbreviated as confinement time) will also trigger the nuclear fusion reaction (Fig. 8).

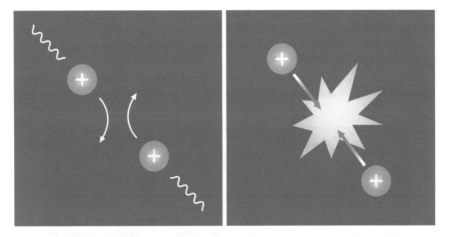

Fig. 6 Higher temperature (faster speed) makes the nuclear fusion reaction more probable

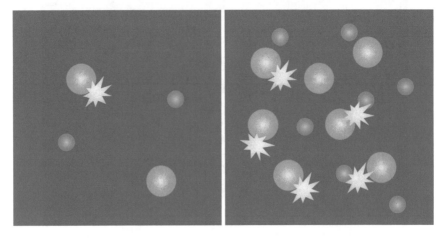

Fig. 7 Higher density (or larger quantity) will increase the probability of the nuclear fusion reaction

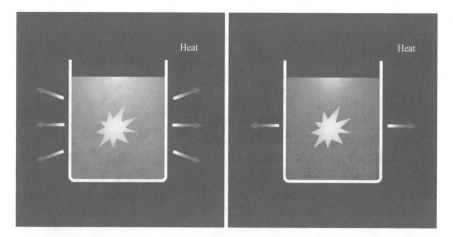

Fig. 8 Prolonging the confinement time will make the nuclear fusion reaction more likely

Thus we can the draw the conclusion: either increasing the temperature or the density, or prolonging the confinement time, the probability of the nuclear fusion reaction will rise. This is the principle of Lawson criterion.

Understanding this condition for ignition, scientists find the direction for future research. They will achieve ignition of nuclear fusion as long as the product of temperature, density and confinement time meets the value calculated by Lawson criterion.

3 All Roads Lead to Rome

After scientists understand how to ignite the nuclear fusion fuel, the nuclear fusion experimental facility development follows two directions: one is focused on increasing temperature of the nuclear fusion fuel until it meets Lawson criterion; the other is to improve the density by squeezing space of the fuel so that it can reach a high value in a short time to meet Lawson criterion.

In Lawson criterion, in addition to temperature and density, there is also confinement time. The nuclear fusion fuel is gas which needs an external force to constrain it to a fixed point instead of dispersing all around so as to make the reaction happen. Scientists call this kind of constraint "confinement". Nuclear fusion reactions generally fall into three types according to different confinement forms (Fig. 9).

The first type is the gravitational confinement nuclear fusion represented by the sun. It mainly relies on gravitation to obtain the confinement force for nuclear fusion gas (primarily hydrogen). The temperature of the sun at the core stands at 15 million °C. Meanwhile, because of gravitation, hydrogen in the surface layer keeps squeezing towards the core to form high density. The high density along with the long confinement time of the sun triggers nuclear fusion reactions naturally.

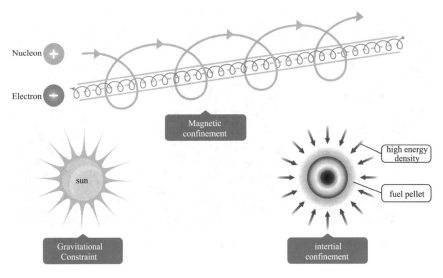

Fig. 9 Three forms of nuclear fusion confinement

The second is the magnetic confinement nuclear fusion based on high temperature. Because equipment for this kind of nuclear fusion needs to be very high-temperature and it's impossible to find materials to directly "contain" the nuclear fusion fuel, scientists have to find another way, i.e. using magnetic field to confine nuclear fusion fuels. This "magnetic container" is suspended in a vacuum chamber to isolate high-temperature fuels from reactors so that the earth can bear fuels with temperature higher than the sun. Because of the high-temperature fuel, this kind of nuclear fusion is called "thermonuclear fusion" (Figs. 10 and 11).

The third type remains inertial-confinement nuclear fusion based on increase of density. This type of equipment usually involves multi-channel and strong laser bombarding simultaneously the ball made of nuclear fusion fuels. Due to the existence of inertia, there is no time for the ball to expand in an extremely short time, and the huge power of centrifugal condensation compresses the nuclear fuel to a high-temperature and high-density state. A short nuclear fusion reaction occurs as a result. The process is completed almost in a second, similar to the process of hydrogen bomb explosion. In fact, hydrogen bomb is also based on the inertia-confined nuclear fusion reaction (Fig. 12).

Among the three types of confinement, the first can't be achieved on the earth. The third, i.e. inertia confinement, can't achieve continuous nuclear fusion reaction because the fuel can't be sustained. The third type is mainly for military use instead of energy production. The only one left ultimately remains the magnetic confinement fusion.

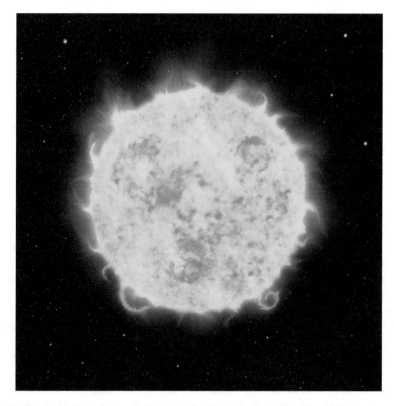

Fig. 10 Gravitational confinement means that the fusion fuel is confined by gravitation

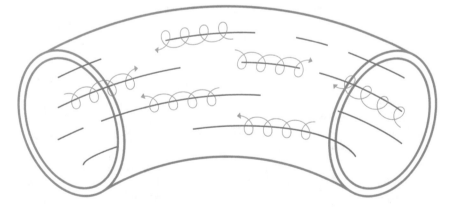

Fig. 11 Magnetic confinement means that the fusion fuel is confined by magnetic fields

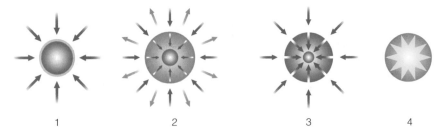

Fig. 12 Inertial confinement means the fusion fuel is confined by inertia

4 Clever Use of the Fourth State of Matter

The reason why magnetic fields can confine the nuclear fusion reaction fuel is that the fuel displays a peculiar state in high temperature—plasma. The state of matter mostly known to people mainly includes three: solid, liquid and gaseous. Plasma does not belong to any of these; it is similar to gas, yet charged internally. Scientists therefore call it the fourth state of matter.

To understand what plasma state is, we need to know something about the micro world. As we know, atoms are made of nuclei and electrons, negatively charged and positively charged respectively. Because the positive and the negative attract each other, the charge is like an invisible kite line that "fastens" electrons so that they always rotate around nuclei and can't "escape". When temperature rises, electrons will rotate faster. And when temperature rises to a certain value, electrons will break free of the invisible lie that "fastens" it and fly freely. On one side are electrons that fly freely; on the other side are nuclei that move freely. The two sides are free from mutual influence and coexist peacefully. At the same time, because they are originally separated from electrically neutral atoms, electrons and nuclei carry the same amount of charge. This state of matter with equal "number of ions" is called plasma (Fig. 13).

> **Knowledge Link**
> **Plasmas in the universe**
> Plasmas are widely scattered in the universe. According to Saha, an Indian scientist, 99% of the visible matter (excluding dark matter) in the universe is the plasma. Red-hot stars, splendid gaseous nebulae, the endlessly extensive interstellar matter and variable ionosphere etc. are all full of plasmas. Plasmas are also common on earth and found in flames, lightning, aurorae and neon lights that are familiar to us (Figs. 14, 15, 16, 17 and 18).

To put it simply, the plasma is a kind of charged gas. Charged matter can be controlled by magnetic fields. Generally speaking, all matter will turn into plasmas when the temperature rises up to 10,000 °C. The temperature of nuclear fusion is

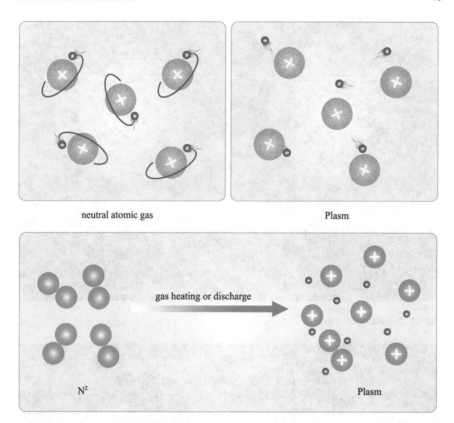

Fig. 13 Heating turns gas into plasmas

usually about 10,000 °C under which condition plasmas are born naturally. This is the very reason why magnetic field can confine nuclear fusion fuels (Fig. 19).

Nuclear fusion fuels will completely turn into plasmas when the temperature is extremely high. If a magnetic field is employed, movement of the plasma will be mainly affected by the electromagnetic (especially magnetic) force. Electromagnetic force, like gravitation, is a kind of non-contact force. From a micro perspective, charged particles will be affected by Lorentz force when they move in a magnetic field. The Lorentz force makes the charged particles to move circularly in the direction perpendicular to the magnetic field. If we employ a very strong magnetic field, radius of the plasma motion will be reduced greatly, its motion trajectory will be compressed and it will be "fastened" to the magnetic induction line.

Fig. 14 Nebula in the universe

Fig. 15 Flames

Fig. 16 Lightning

Fig. 17 Aurorae

Fig. 18 Neon lights

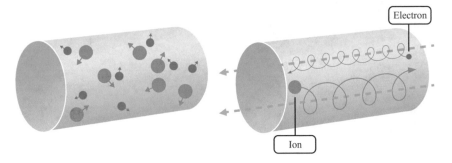

Fig. 19 Diagram of the magnetic field confining plasmas

Knowledge Link

Lorentz Force

The Lorentz Force is the force exerted on a charged particle moving in the magnetic field. It was a basic hypothesis proposed by the Dutch physicist Hendrik A. Lorentz in 1895 when he developed the classic "Theory of Electrons". The hypothesis was proved later on in many experiments and named after him. According to the International System of Units, the unit of Lorentz force is Newton, the symbol for which is N. The Lorentz force is always perpendicular to the motion direction of charged particles. The Lorentz force does not change the velocity and kinetic energy of charged particles. It only changes their motion direction and causes deflection.

5   Magnetic Field Flattened and Made into Torus

The temperature of the nuclear fusion fuel is very high and yet the most heat-resistant material on the earth can only bear temperature of several thousands of degrees Celsius. No actual material can be found to "withstand" these fuels. Now we have a help hand from the magnetic field, "withstanding" the scorching nuclear fusion fuel becomes a reality.

Scientists use the very strong magnetic field to make a "cage" and then put the red-hot nuclear fusion fuels into the "magnetic cage". Finally, the "cage" and the actual material are isolated by a vacuum. In so doing, the earth "withstands" an artificial sun whose temperature is higher than the sun (Fig. 20).

Nevertheless, as something invisible, the magnetic field can't be controlled as easily as other visible materials and suffer uncertainties. This brings huge challenge for us to successfully "withstand" the scorching nuclear fusion fuels. To address the problem, scientists think of many ways to "flatten" the field and then "knead" it into a circular form so as to find the most suitable and most stable configuration for the magnetic field.

Up till now, scientists have designed various forms for the magnetic field and developed many magnetic confinement facilities with different forms for nuclear

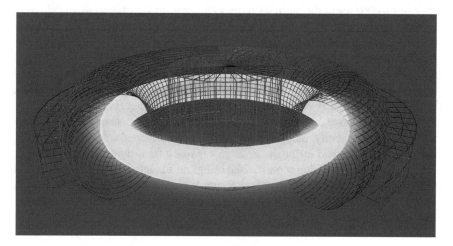

Fig. 20   The "cage" made of the magnetic field can "be filled with" high-temperature nuclear fusion fuels

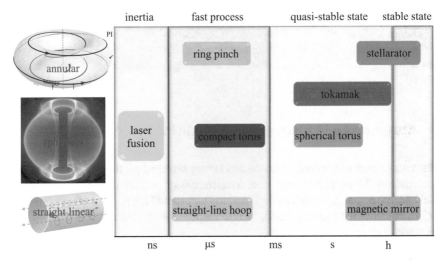

Fig. 21 Simple categorization of magnetic confinement facilities

fusion reactions. Some are shaped like sweets, some like fried-dough twists and some like doughnuts etc.; the fields in various forms carry their own pros and cons. Here are some simple examples (Fig. 21).

The lateral axis is the confinement time; the vertical axis is the configuration. Laser fusion is used as a reference value (图说明文字).

The first is the magnetic mirror. The magnetic mirror is a weak magnetic field with strong ends. It looks like a sweet. When moving circularly around the magnetic induction line from or to the ends, charged particles will be exerted a reverse force. This force forces the particles to slow down until they stop moving forward and bounce off. After forced back, the particles move circularly through the weak middle field towards the other end and bounce off again because of the reverse force. The plasma is ultimately confined in such magnetic field configuration through the bouncing-back-and-forth process. Because the process is like light travelling between two mirrors, the facility is called magnetic mirror. This magnetic mirror is a simply configured magnetic field. The advantage of simplicity is that it's easy to produce and convenient for research. The flaw is that the confinement is not so strong (Fig. 22).

In order to improve the confinement effect, scientists design another more sophisticated magnetic facility. Stellarator is a typical sophisticated magnetic configuration. The magnetic field inside the stellarator is like a fried-dough twist, something seemingly confusing. Being sophisticated brings an advantage of good confinement effect to the point that the nuclear fusion fuel is constrained very stably inside the reactor. Yet the flaw is that the research, design and construction are very challenging (Fig. 23).

(a)

(b)

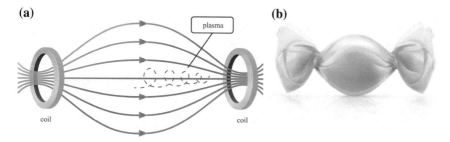

Fig. 22 The configuration of the magnetic mirror looks like a sweet

(a)

(b)

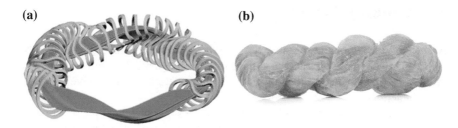

Fig. 23 The magnetic configuration of Stellarator looks like a fired-dough twist

Being too simple means weak confinement and being too sophisticated means challenging construction. Scientists then design another magnetic field configuration—tokamak. Tokamak magnetic field is torus-shaped, somewhat like a doughnut. The advantage of this configuration is that it forms a closed loop and that it is not so sophisticated. Closed loop allows charged particles to move in circulation with better confinement effect. Being not so sophisticated means that it is easy to be developed and researched. At present, tokamak has the best effects among all nuclear fusion facilities. It is therefore the most frequently researched and most mainstreaming magnetic configuration in the world (Fig. 24).

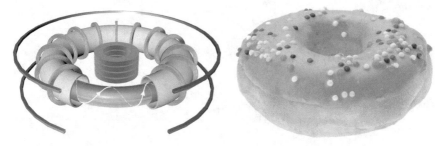

Fig. 24 The magnetic configuration of the tokamak is like a doughnut

6   The Mighty Doughnut

At the core of the tokamak facility lies a toroidal vacuum chamber surrounded by coils. When the facility is charged, a huge toroidal magnetic field will be generated inside to heat the nuclear fusion fuel to very high temperature so as to achieve nuclear fusion reactions.

The internal magnetic field of tokamak is like a super doughnut placed horizontally. In order to create a magnetic field in the shape of such doughnut, tokamak is installed with three types of fields: the central solenoid is used to stimulate plasmas to generate toroidal plasma currents; the toroidal field is employed to generate toroidal magnetic field to confine plasmas so that they rotate around the toroidal field; the outer poloidal field is used for controlling location and configuration of plasmas (Fig. 25).

The three tokamak fields perform their duties and generate a closed toroidal magnetic field under command of the control system to stimulate, confine and control

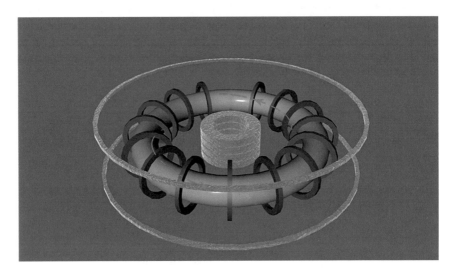

Fig. 25 The magnetic field of Tokamak

plasmas. Due to the smart design and the advantage that the construction process is not as sophisticated as that of the stellarator, the tokamak becomes a nuclear confinement focus rapidly in nuclear fusion research. There is a general agreement that tokamak is the most promising facility to realize nuclear confinement fusion.

Knowledge Link
Fields in Tokamak
    Tokamak is mainly composed of the Central Solenoid, the Outer Poloidal Field and the Toroidal Field. The vertically installed Toroidal Field is also called the manipulation field and the horizontally placed Central Solenoid and Outer Poloidal Field are jointly called poloidal fields. In actual installation, the number of fields of each type differs based on the tokamak design. Apart from the three major fields, some tokamaks are also designed with small fields to optimize the magnetic configuration.

    The history of magnetic confinement research for nuclear fusion reactions is also a history of tokamak development. Over the past half century, scientists have built over 100 tokamak facilities, big or small. During the exploration process of magnetic confinement-based nuclear fusion reactions by humanity, why did tokamak gain popularity overnight? What are the proud achievements made during the development process? Why tokamak is considered the most promising facility to realize nuclear fusion? What is the future of tokamak? And what are the difficulties in developing tokamak? Our next chapter will unveil the answers.

The Past and Present of Tokamak

Teng Wang

Abstract As the magnetic confinement facility for nuclear fusion experiments, tokamak is easy to be developed with sound confinement effects. It is thus considered the most promising facility to realize magnetic confinement fusion. This chapter begins with the origination of tokamak and goes on with its history in a chronological order.

1 The History of Tokamak

Tokamak has a relatively long history. Six decades have passed since 1950s when Soviet scientists proposed the concept of tokamak for the first time. During this period, scientist have made remarkable achievements and also conquered one mission seemingly impossible after another.

In 1950s, Soviet scientists built the first tokamak facility in the world and named it "T-1". The coming decades ever since have been the exploration stage of tokamak. After continuous improvement, scientists finally achieved desirable experiment results in facilities represented by T-3 and won recognition from countries around the world. Therefore, tokamaks have gradually become the optimal option for nuclear fusion research (Fig. 1).

In 1980s, countries in the world designed and developed a large number of tokamak facilities. Through continuous experiments and improvements, experiment parameters from tokamak have been constantly updated. Tokamak has gradually come to the spotlight of magnetic confinement fusion research.

Based on previous research, scientists found that "the bigger the facility is, the better the parameter will be". Several countries which researched nuclear fusion at that time began to build large-scale tokamak facilities. The Europe built JET (Joint European Torus); the US built TFTR (Tokamak Fusion Test Reactor); Japan built JT-60 (Japan Torus-60). In 1980s and 90s, JET and TFTR achieved deuterium and tritium nuclear fusion reactions, yet with only a few seconds of duration.

T. Wang (✉)
Institute of Plasma Physics, Chinese Academy of Sciences, Hefei, Anhui, China
e-mail: yihco0101@sina.com

© Zhejiang Education Publishing House 2021
B. Wan (ed.), *Man-Made Sun*, China's Big Science Facilities,
https://doi.org/10.1007/978-981-16-3887-9_5

Fig. 1 Tokamak presents the major approach to nuclear fusion research for countries around the world

Since 1990s, in order to achieve long-time stable operation of tokamaks, scientists started to build superconducting facilities. China became the first to build the superconducting tokamak facility—EAST. The nuclear fusion reactor ITER jointly designed by multiple countries is also under joint construction. The superconducting tokamak keeps updating the operation records. Stable operating duration has been prolonged from the initial several seconds to dozens of seconds and hundreds of seconds (Figs. 2 and 3).

Up till now, scientists in the world have built over 100 tokamak facilities, big or small. They all make their respective contribution to fusion research. Next, we will introduce several milestone tokamak facilities represented by T-3, JET, TFTR, JT-60 and ITER.

2 A Household Name Overnight

Tokamak was born in the Soviet Union. In the early 1950s, the Soviet scientist Tamm and Sakharov proposed "tokamak" for the first time.

In 1957, Sector-44 from the Moscow Measurement Instrument Science Laboratory (predecessor of the Kurchatov Institute) started tokamak research. The first tokamak facility in the world was developed by Artsimovich etc. and was put into operation in 1958 after completion.

In the beginning, tokamak didn't attract much attention. While Soviet scientists researched tokamak, developed countries in the west also tried to develop nuclear fusion facilities of various magnetic fields such as stellarator to achieve magnetic confinement fusion and yet only suffered disappointment.

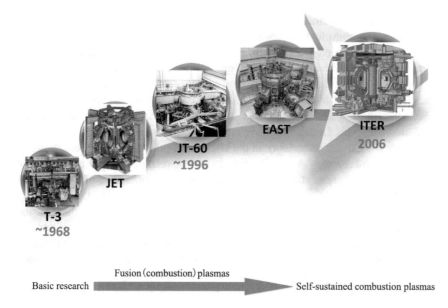

Fig. 2 From T-3 to ITER—major tokamak facilities in history

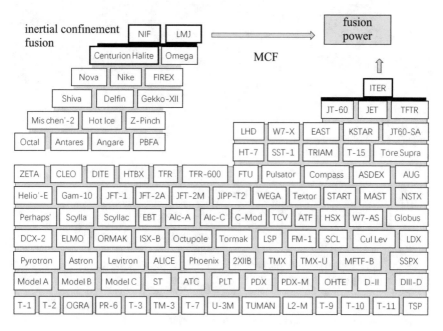

Fig. 3 List of tokamak facilities in history

Until August 1968, International Atomic Energy Agency (IAEA) Fusion Energy Conference was convened in Novosibirsk, Soviet Union. During the conference, Artsimovich announced latest achievements they had made. Good experimental results continued in the Soviet Union in developing T-3 tokamak: electron temperature reached 1,000 °C and plasma temperature was as high as 5 million °C. These parameters were 10 folds higher than those of the Princeton stellarator (Fig. 4).

Scientists didn't believe that the Soviet facility outperformed theirs so much and questioned the result. In order to prove authenticity, Artsimovich invited a research group from the Culham laboratory in the UK to bring the most advanced Thomson scattering measurement system at that time to test the result at Kurchatov Institute in Moscow. The measurement results show that electron temperature of T-3 tokamak is (almost one fold) higher than the 1,000 °C Artsimovich announced.

The result caused an unprecedented sensation in magnetic confinement fusion the moment it was released. Tokamak achieved a historic breakthrough and the sensational T-3 also made history in the nuclear fusion research. Scientists across the world started to carry out research on tokamak. Since then, tokamak magnetic confinement fusion research embraced a new era. Tokamaks big and small have been mushrooming (Fig. 5).

Fig. 4 T-3 impressed the IAEA fusion energy conference in August 1968

Fig. 5 Western scientists tested the temperature of T-3 and found that it was higher than the result announced by the researcher

3 The Imposing Four Guardians

After huge success of T-3 tokamak, a sensation spread across the world. Countries started to build or upgrade to large-scale tokamaks. From 1970s to the mid 1980s, four large-scale tokamaks had been built: JET in the Europe, TFTR in the US, JT-60 in Japan and T-15 in the Soviet Union. Apart from T-15 which failed to operate due to collapse of the Soviet Union, the restall obtained major achievements (Fig. 6).

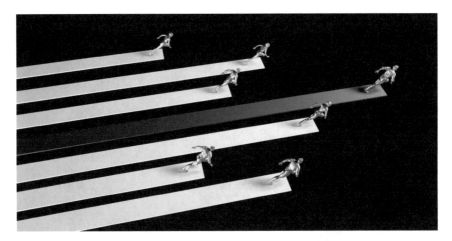

Fig. 6 The success of T-3 sparked a competition in tokamak development across the world

3.1 The Pride of Europe—JET Tokamak

In 1980s, Europe aimed to build a large-scale experimental facility to study the fusion physics of deuterium–tritium fuels and maintain and repair the facility through tele-control. Then JET was born against such backdrop.

JET is a flagship in Europe's nuclear fusion plan. JET is admiring from the perspective of either science and technology or scientific management. This a large-scale science project under cooperation of top European scientists, industrial technicians and innovative management team. JET's concept and key features are quite different from those of other large-scale tokamaks designed in the late 1970s and early 1980s (Fig. 7).

The uniqueness of JET demonstrates in the D-shaped toroidal fields, vacuum chamber and large-volume high-current plasmas. The advanced JET facility remains a desirable achievement in science and technology. In 1997, the deuterium–tritium experiment (DTE1) at JET set the world record for producing 16 Mw transient fusion power. Meanwhile, JET's output power exceeded the input power (with the Fusion Energy Gain Factor $Q = 1.25$), a success earned for the very first time in the history of humanity. The result also proved through experiment that the Fusion Energy Gain Factor Q is a pivotal parameter for magnetic confinement fusion in addition to the reaction temperature (T), reactant density (n) and confinement time (τ). The fact that JET facility achieved $Q = 1.25$ not only proves that tokamaks have the possibility of net output power but also shows the feasibility of that possibility. This is recognition of tokamak confinement fusion.

Fig. 7 JET tokamak facility

Knowledge Link
Fusion Energy Gain Factor
The Fusion Energy Gain Factor Q = nuclear fusion output power ÷ power input in a nuclear fusion facility. Because the fusion facility is hugely energy-consuming, the input (consumed) power is enormous and the facility has to ensure the "Q > 1" result to make it valuable.

3.2 Attempt of the US—TFTR Tokamak

TFTR is a large-scale tokamak experimental facility built by the Princeton Plasma Physics Laboratory in the US at the cost of $314 million. The facility was completed and put into operation in 1982 (Fig. 8).

TFTR facility was initially built to realize the balance between input fusion power and output power. Although it failed to realize this goal, TFTR facility made major progress in exploring and understanding features of plasmas such as deuterium and

Fig. 8 TFTR in the US

tritium. The most significant step was achieved on December 9th and 10th 1993 when the composite fuel made of 50% of deuterium and 50% of tritium reached temperature of 500 million °C. The peak fusion power of plasmas such as deuterium and tritium reached 10.7 MW.

From 1982 to 1997, TFTR facility had made huge contribution to nuclear fusion reactions in terms of confinement time and energy density. TFTR also provided information on confinement, heat and α-particle physics and experience in handling deuterium and activating DT neutrons in experimental environment. The information offered valuable reference for later design and construction. Due to some reasons, TFTR was shut down in 1997.

3.3   The Success of Japan—JT-60 Tokamak

JT-60 is an experimental facility aimed at achieving critical state for plasmas (power gain factor Q > 1.0).It is considered one of the equally important top three tokamaks along with JET and TFTR. The facility began operation on April 4th 1985; the total cost was 23 million JPY (about 15.3 billion RMB). Its main objectives are: to reach critical state of plasmas; to confirm confinement law under plasma state,

Fig. 9 JT-60 in Japan

secondary heating and impurity control. Deuterium-deuterium (D-D) reaction test was conducted successfully at JT-60 facility. Converted to D-T reaction, the gain factor Q could reach 1.00. After that, the Q value exceeded 1.25 (Fig. 9).

JT-60 was upgraded to JT-60U from 1989 to 1991. Tests were carried out later on to improve confinement performance and stable operation. The objective was to study stable operation of tokamak facility by improving confinement effects of plasmas. The gain factor exceeded 1.3 at JT-60U facility, a result calculated based on the D-D tests.

Knowledge Link

JT-60

JT-60 is not only a representative facility of Japan's magnetic confinement fusion project, but also a star in the world's magnetic confinement fusion research. It was initially operated by Japan Atomic Energy Research Institute (JAERI) and currently operated by Naka Fusion Institute of Japan. Since its operation in 1985, JT-60 has been keeping the highest records in the world of the triple product (plasma temperature, density and confinement time which directly show the possibility of commercial use of nuclear fusion energy).

JT-60U (Japan Torus-60 Upgrade) is the upgrading version of JT-60. After the upgrading, JT-60U had higher plasma confinement performance in fusion reactions and attained sound results in stable operation tests. JT-60U stopped operation in 2008 and was transformed to a superconducting tokamak.

JT-60SA (Japan Torus-60 Super Advanced) will be a superconducting tokamak. Drawing on JT-60U, the new facility has been completely transformed with changed plasma configuration, upgraded plasma volume and new superconducting magnets. The project will be carried out under cooperation of Japan and the European Union. The facility is under construction and expected to be completed in 2019.

3.4   The Regret of the Soviet Union—T-15 Tokamak

T-15 facility was from Kurchatov Institute in Moscow and its large radius reached 2.4 m. T-15 realized plasma discharge for the first time in 1988. Due to shortage of funds, the facility was shut down in 1995. Despite only 100 times of discharge during its short existence, T-15 still reached 1.5 MW input power and 1 MA electric currents which sustained 1 s of plasma discharge. Despite a flash in the pan, the result is still splendid (Fig. 10).

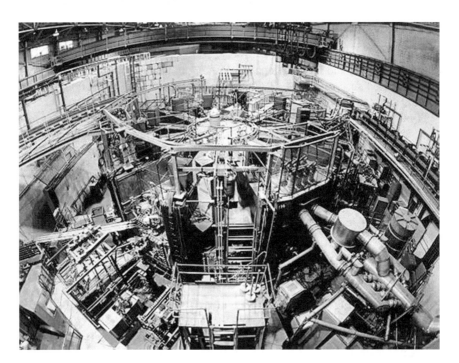

Fig. 10 T-15 of the Soviet Union

4 Superconducting Application to Nuclear Fusion

To meet the requirement of magnetic confinement for high-temperature plasmas, conventional tokamaks use common copper conductors to build huge fields. Although controlled nuclear fusion can be achieved, the facilities suffer flaws of large volume, low efficiency and few breakthroughs. At the end of twentieth century, scientists started to apply superconducting technology to tokamak facility. This is a huge breakthrough in controlled nuclear fusion research. It turns continuous stable operation of magnetic confinement fusion into reality.

The magnetic confinement fusion facility requires strong magnetic field to constrain nuclear fusion fuels, and a strong field requires powerful electric currents. If there is strong resistance on the coils, intense heat will be generated and damage the magnetic confinement facility. Predecessors of T-7 can only achieve short-pulsed operation with a large volume of power consumed. Superconductors have greatly enhanced operation duration of nuclear fusion facilities.

Knowledge Link

Advantages of superconductors

In tokamak, powerful currents up to 10,000 or 100,000 amperes must be applied through the coil to create a magnetic field strong enough. This amount is 1,000 or even 10,000 times of household air-conditioning currents. The common copper wire has resistance, though small. The resistance will cause serious heat problems as a result of powerful currents and thus limit the duration of tokamaks to some extent. Superconducting materials are resistance-free in special conditions. This advantage will help avoid heat generated in the coil and prolong the duration. In addition, compared with the common copper wire, superconductors, with smaller size, deliver the same amount of current, which helps miniaturize and compact the tokamak magnetic system.

T-7 built by the Soviet Union in 1970s is the first superconducting tokamak facility in the world. It had been operated for five years and stopped operation after that. The Soviet Union then focused on a larger superconducting tokamak—T-15. France and Japan also followed the step to build their superconducting tokamak (Fig. 11).

In January 1990, the Soviet Union presented T-7 tokamak to China as a gift. China then started the research of the superconducting tokamak. With the help from Russian experts, China transformed T-7 thoroughly and renamed it HT-7. In 1994, HT-7 was completed. This makes China the fourth country in the world following Russia, France and Japan to possess the superconducting tokamak.

The construction and operation of HT-7 urshed in a new era of nuclear fusion in China. It brings China to the frontier of the research on superconductor, cryogenic refrigeration, strong magnetic field etc. and makes huge contribution to the world's fusion research (Figs. 12 and 13).

Fig. 11 Superconducting strand samples

5   China Pioneers the Fully Superconducting Tokamak

Tokamaks like HT-7 is not fully superconducting. Only part of the fields that carry stable currents are applied with superconducting materials, and the rest still employ common conducting materials. The reason for the partial application is that super conductors are extremely difficult to operate and will hugely increases the development costs. The future fusion reactors must be durable, which means that all coils need to be superconducting. The ultimate goal of tokamak research is to build fusion reactors for commercial use. The super conducting tokamak is hence an irresistible trend (Fig. 14).

To support the long-term development of the nuclear fusion research, the Chinese government allocated a specialized fund to IPP for achieving breakthroughs in this area. The project gathered almost all top nuclear fusion talents in China. The researchers delivered on the mission and developed the first non-circular cross-section superconducting tokamak in the world—EAST.

EAST is the first superconducting tokamak in the world in that all the 30 main internal fields adopts superconducting materials. The project started in 1998 with initial name of HT-7U and competed in 2006. Its engineering commissioning was also completed in the year (Fig. 15).

Fig. 12 T-7 tokamak of the Soviet Union

Success of EAST signifies the coming of a full superconducting era. It also improved China's overall strength infusion research. At present, the world has two fully superconducting tokamaks in operation: China's EAST and South Korea's KSTAR (built in 2009). France and Japan are upgrading their partially superconducting tokamak—Tore Supra and JT-60U—into fully superconducting WEST tokamak and JT-60SA tokamak respectively (Figs. 16, 17 and 18).

Fig. 13 HT-7 superconducting tokamak of China

Fig. 14 Fully superconducting EAST facility

Fig. 15 The toroidal field of EAST is D-shaped (non-circular cross-section)

6 Global Cooperation in ITER

Worldwide exploration over the decades has made tokamak-oriented nuclear fusion research increasingly mature. Before the commercial use is realized, however, some scientific and technological problems need to be tackled in tokamak-based nuclear fusion research and development plan. Against the background that many large and

Fig. 16 Fully superconducting KSTAR facility in South Korea

Fig. 17 Fully superconducting WEST facility in France

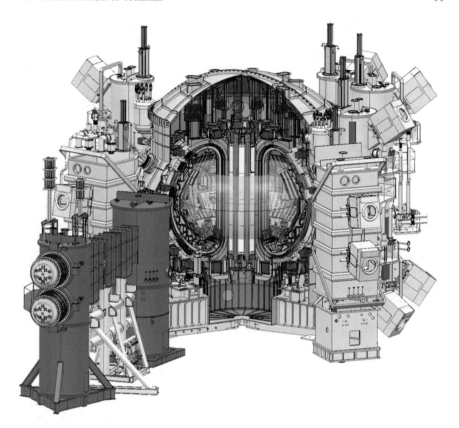

Fig. 18 Fully superconducting JT-60SA facility in Japan

medium-sized tokamaks have already achieved remarkable progress, ITER comes about to verify the capacity of long-time nuclear fusion energy output and address the most critical engineering problems to meet the future demand for highly efficient, compact and stably operating commercial reactors (Fig. 19).

The term "ITER" comes from initials of International, Thermo nuclear, Experimental and Reactor. In 1985, the Soviet leader Gorbachev and the US President Regan proposed in the Geneva Summit that the US, Soviet Union, Europe and Japan would jointly launch the ITER project. ITER is aimed at building a tokamak fusion reactor which enables self-sustained combustion so as to deeply explore physical and engineering issues of future DEMO fusion reactors (DEMO) and commercial fusion reactors.

ITER was jointly developed by seven parties, i.e. China, EU, Japan, South Korea, Russia, the US and India. China officially joined the project in 2003. The members cover almost all nuclear powers in the world and their populations account for over half of the world's total. ITER's ultimate goal is to achieve stable state; to prove

Fig. 19 The whole world begins to cooperate in the development of nuclear fusion energy

controlled ignition and sustained combustion of D-T plasmas; to verify key technologies related to reactors through comprehensive fusion systems; to conduct comprehensive tests on high heat flux and nuclear radiation components for peaceful use of nuclear fusion energy. Research findings and experience will help to build and enable smooth operation of the DEMO. And success in the DEMO will make commercial application of the nuclear fusion energy in the future possible.

Knowledge Link

DEMO

DEMO means demonstration. Here it refers to the device between ITER and commercial fusion reactors that verifies the feasibility of nuclear fusion power plants. In this sense, DEMO also means the fusion demonstration reactor. DEMO will draw on successful operation of ITER and focus on three objectives. The first is to realize high net power output; the second is to achieve the proliferation of tritium as the nuclear fusion material; the third is to verify that all technologies required for building the commercial fusion power plant can be obtained through commercial operations. Therefore, the focus of tokamak development following ITER will be stable state, self-sustained tritium, energy exchange and high net power output. Meanwhile, tokamaks will come out of the laboratories for industrial and commercial use as large-scale fusion reactors. Therefore, ITER project is an important milestone in the research and utilization of nuclear fusion energy and a critical step in human history from research on controlled nuclear fusion to commercial use.

ITER project consists of three stages: the first stage is for construction of the experimental reactor from 2007 to 2025; the second stage is for experimental operation of thermonuclear fusion in 20 years, during which scientific and technological verification for large-scale commercial development of nuclear fusion energy will be conducted on performance of nuclear fusion fuels, reliability of materials used for reactors and development feasibility of fusion reactors; the third stage is for retirement of the experimental reactor in five years (Fig. 20).

ITER facility integrates not only latest results in international fusion research, but also cutting-edge technologies in related fields available up till today, including large-scale superconducting magnet technology, the medium-energy high-intensity accelerator, continuous high-power microwave technology, sophisticated remote control,

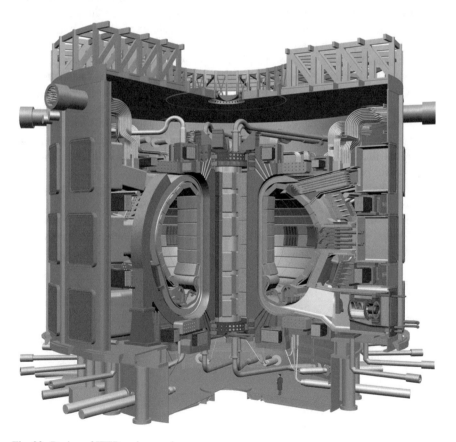

Fig. 20 Design of ITER main vessel

reactor materials, experimental claddings, large-scale cryogenic technology, tritium technology, advanced diagnosis technology, large-scale power source technology and fusion safety etc. These technologies are not only necessary for future nuclear fusion power plants, but also will make huge contribution to industrial, social and economic development of countries all over the world. The development, operation and experimental research of ITER is a necessary step for humanity to explore fusion energy and it will be likely to directly impact the design and development of DEMO and commercial use of fusion energy—the fusion power plant.

7   Roadmap for Fusion Development

Since its birth, tokamak has been the "favorite" of the nuclear fusion research. To some extent, tokamak has become a representative of the fusion research. Thanks to hard work of generations of researchers, tokamak research has developed from isolation to cooperation, from the conventional to the superconducting and from the small-sized experimental facility to the large-scale experimental reactor. Yet, humanity's expectation is still high to be met.

To embrace the day for peaceful utilization as early as possible, the seven parties of ITER have disclosed or partially disclosed their roadmap for nuclear fusion development. On the basis of the current tokamak facilities and ITER experimental reactors under construction, all parties agree that there will be a stage for the DEMO reactor before the construction of commercial fusion power plants. The DEMO reactor is aimed at ensuring that all problems will be solved and that all procedures meet standards in the final verification process. Only under this condition will the commercial fusion power plant be realized (Fig. 21).

Tokamaks are constantly upgraded with continuous hard work of researchers. It is likely that we energy from nuclear fusion power plants will be available in the second half of the twenty-first century. The very "heart" of the fusion plant is the "lead" of this chapter—tokamak. By that time, we will bid a farewell to energy crisis and environmental pollution.

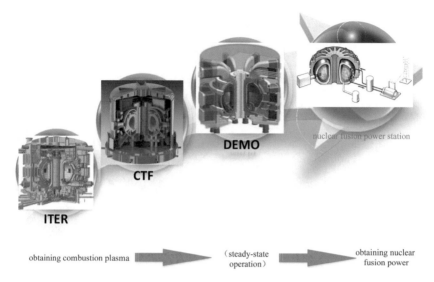

obtaining combustion plasma　　　　　　（steady-state operation）　　　　　　obtaining nuclear fusion power

Fig. 21 Possible roadmaps for future tokamak research

The Road of China to Artificial Sun

Teng Wang

Abstract China's research on controlled nuclear fusion dates back to the late 1950s. Institute of Plasma Physics (IPP) of Chinese Academy of Sciences and Southwest Institute of Physics (SIP) under China National Nuclear Corporation (CNNC) were founded for specialized research of the field. At present, the research in China is conducted mainly with the tokamak facility.

1 Start from Scratch: The Hardship of Startup

In 1968, the Soviet T-3 tokamak facility shocked the world with its plasma parameters and stability. Since then, nuclear fusion research embraced the tokamak era.

Chen Chunxian, who worked in Institute of Physics, Chinese Academy of Sciences at that time, noticed the remarkable achievement in science community and realized the long-term significance of tokamak. In around 1970, Chen proposed to conduct tokamak research for many times. In 1972, Institute of Physics of CAS formed tokamak research group and Chen was appointed as the head. The group began to develop China's first tokamak facility with iron-core transformer—Beijing tokamak 6 (CT-6) which was completed in 1974 (Fig. 1).

In 1973, CAS decided to build "Hefei Experimental Station of Controlled Thermonuclear Fusion Research" in Hefei City (abbreviated as "Hefei Controlled Station"). In 1978, CAS officially set up Institute of Plasma Physics (abbreviated as IPP), and appointed Chen etc. to be the leadership of the institute.

After the founding of the IPP, the 6th facility of Hefei Controlled Station was transformed. In 1980, the transformation was completed and the facility was renamed HT-6A. After over one year of operation, research progress was obtained in balance commissioning, basic diagnosis and disruptive instability.

In 1982, IPP upgraded HT-6A again. Main components such as toroidal fields and the vacuum chamber were reprocessed. The upgrading was completed in 1983 and the facility was renamed HT-6B. In 1993, HT-6B stopped operation (Fig. 2).

T. Wang (✉)
Institute of Plasma Physics, Chinese Academy of Sciences, Hefei, Anhui, China
e-mail: yihco0101@sina.com

© Zhejiang Education Publishing House 2021
B. Wan (ed.), *Man-Made Sun*, China's Big Science Facilities,
https://doi.org/10.1007/978-981-16-3887-9_6

Fig. 1 Chen Chunxian (1934–2004), one of the founders of Institute of Physics, Chinese Academy of Sciences

While upgrading the 6th facility, IPP launched another project in 1980 to design a bigger tokamak facility and completed the project in 1984. The new facility is named HT-6M. In 1986, CAS held an inspection and evaluation meeting for HT-6M.Evaluation experts all agreed that HT-6M facility has excellent performance and meets international standards for facilities of that type. The success means that IPP has possessed the capacity for developing the medium-sized controlled nuclear fusion facility (Figs. 3 and 4).

Meanwhile, another important base for controlled nuclear fusion research in China—the CNNC Southwest Institute of Physics located in Chengdu also obtained major progress in tokamak facility research and development—China Circulator 1(HL-1). In 1995, HL-1 was upgraded to HL-1M facility with the thick copper shell and the internal toroidal fields were removed. Main plasma parameters witnessed obvious improvement as a result. In 2002, SIP of CNNC transformed the divertor-based ASDEX tokamak facility given by Germany into a large-scale tokamak experimental facility with divertor configuration—China Circulator 2(HL-2A). The facility had been tested since 2002 and became the first in China to achieve the discharge with divertor configuration (Fig. 5).

From the first tokamak CT-6 to the most advanced conventional tokamak HL-2M (upgrading version of HL-2A) at present, China's conventional superconducting tokamak has developed from the small to large size and from low parameters to high parameters. Besides, apart from professional institutes, some domestic universities

Fig. 2 HT-6B tokamak facility

also began to conduct tokamak research. Examples include the KT-5 small-sized facility developed by University of Science and Technology of China in 1984, China's first small-sized low-aspect ratio tokamak—SUNIST developed by Tsinghua University by the end of 2002, J-TEXT transformed and renamed by Huazhong University of Science & Technology from the original TEXT-U given by the US etc. These facilities are used not just for research, but more importantly for cultivating research talents in controlled nuclear fusion.

2   Grasp Opportunities and Conquer Challenges

At the end of twentieth century, scientists started to apply latest superconducting technology to tokamak facilities. The first superconducting tokamak facility in the

Fig. 3 HT-6M tokamak facility

Fig. 4 CAS held the Inspection and Evaluation Meeting for HT-6M

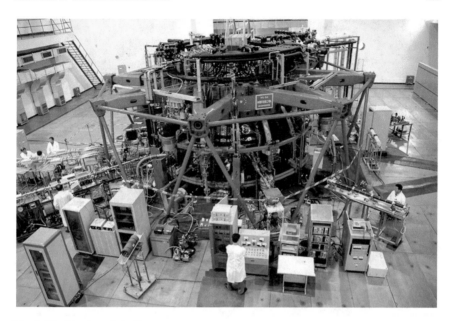

Fig. 5 HL-2A tokamak facility

world—T-7 was developed by the Soviet Union in 1979. Soon after that, the Soviet Union built a bigger one—T-15. T-7 was then left unused after five years of operation at the Kurchatov Institute and tests of the facility basically stopped after 1985. Later on, academician Kurchatov, Director of the Kurchatov Institute wrote to China that they were willing to give the facility to China. The letter was handed over to academician Huo Yuping, the then Director of the IPP (Figs. 6 and 7).

At that time, IPP already had a relatively large non-superconducting tokamak project and completed the engineering design. Director Huo Yuping analyzed thoroughly the international trend in nuclear fusion research and believed that superconducting magnetic confinement fusion facility was an inevitable trend. In addition, China was not able to manufacture some key components of T-7 at that time. IPP, therefore, decided to grasp the opportunity to introduce T-7 and halted the ongoing non-superconducting tokamak project. Introduced to China, T-7 was then transformed into a sound nuclear fusion facility. This is a very forward-looking decision from present perspective.

In October 1990, IPP reached an agreement with the Soviet Union that T-7 facility would be introduced through barter trade. In the following three years, components and relevant parts of T-7 were transported by a total of 47 train carriages to Hefei. T-7 only has 12 small windows and can't be used for the plasma physics experiment in real sense. IPP carried out a fundamental transformation on the facility with all major human and relevant resources available: the original 48 toroidal fields were combined and changed to 24; the new vacuum chamber was designed and developed; 34 new windows were added. All these greatly facilitated access to the facility. In order to

Fig. 6 Academician Kurchatov, Director of the Kurchatov Institute

conduct experiments on high-power assisted heating and long-pulsed operation, the facility was designed with active water-cooling lining in the vacuum chamber and new toroidal fields. Nine subsystems were developed, including the largest cryogenic liquid helium system in China and high-power system. The original T-7 facility that couldn't allow for physics experiments was transformed into an advanced facility that enables various types of experiments (Fig. 8).

After transformation, T-7 facility was renamed HT-7 of which H means Hefei. Unlike other conventional tokamaks, HT-7 adopts niobium and titanium (Nb-Ti) superconductor for the toroidal fields. Because the superconductor is resistance-free, the coils will not generate heat even if a large volume of current passes through. HT-7 is medium-scale nuclear fusion facility for research which produces long-pulsed and high-temperature plasmas.

Commissioning of HT-7 was completed in 1994. One year later, the facility was officially put into operation. HT-7 costed only 200 million RMB for installation and supporting devices. One tenth of the budgetary investment gave birth to an advanced, by the standard at that time, large-scale pulsed superconducting nuclear fusion research system. Since then, China has been approaching fast towards the frontier of the international nuclear fusion research (Fig. 9).

HT-7 has operated for 18 years during which nearly 20 scientific experiments were conducted with over 100,000 times of electric discharge. Major achievements were earned from engineering and physics perspectives. In particular, an experiment carried out on March 31st, 2003 achieved over one minute of plasma discharge. This

Fig. 7 The T-7 Superconducting tokamak facility

marks the second tokamak facility after Tore Supra facility in France to enable high-temperature plasma discharge at minute level. It also signifies that China enters a new stage of nuclear fusion research; China's power in the research has been significantly improved (Fig. 10).

IPP continued to upgrade the facility. It improved the first high-performance water-cooling wall and feedback on the precise real-time long-pulse plasma location and on the plasma density etc. On March 15th 2008, the staff of IPP witnessed the 100,000th discharge of HT-7 superconducting tokamak. At midnight of March

Fig. 8 Rebuilding HT-7 facility in 1993

Fig. 9 Main vessel of the HT-7 superconducting tokamak

Fig. 10 Researchers witness the 100,000th discharge from the HT-7 superconducting tokamak facility

12th, 2008, HT-7 superconducting tokamak hit another record in physics experiment: it achieved plasma discharge for 400 s continuously with the electric temperature reaching 12 million degrees Celsius. This is the longest high-temperature plasma discharge among facilities of the same type in the world at that time (Fig. 11).

During its operation for nearly 20 years, HT-7 realized all objectives set in the beginning and some results even surpassed expectations. On October 12th 2012, HT-7 conducted the last discharge experiment which marksan ending of its "career". After that, the facility officially "retired" and became one of the big science project facilities that received approval of "retirement" (Fig. 12).

HT-7 is not only the first superconducting tokamak in China, but also one of the few superconducting tokamaks in the world. Its successful development and operation enabled China to arrive at the frontier of the international nuclear fusion research explore an innovation-driven path for the country to come to the global stage of the nuclear fusion field in an all-round manner.

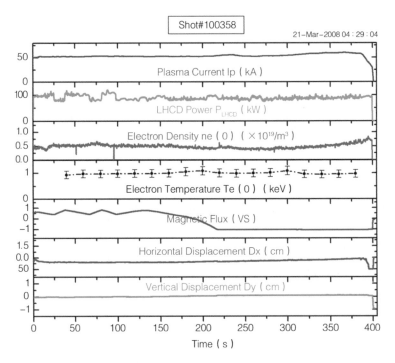

Fig. 11 HT-7 achieved plasma discharge for 400 s

Fig. 12 Main vessel of the "retired" HT-7 tokamak

3 The Latecomer Becomes the Pioneer

After the success of HT-7 superconducting tokamak, IPP proposed to develop a fully superconducting tokamak. Back to the time, only part of the toroidal field of the existing superconducting tokamaks adopted superconducting materials. There was no precursor of the fully superconducting tokamak in the world due to high cost and difficulties in development. Nevertheless, future nuclear fusion power plants require long time of stable operation, which means that all magnets need to be superconducting. Evolve from the conventional to the superconducting tokamak presents the trend in the research. It is therefore imperative to build the fully superconducting tokamak (Fig. 13).

On the basis of the successful superconducting HT-7 tokamak, IPP proposed the plan of "building HT-7U—a fully superconducting non-circular cross-section tokamak facility". The facility was renamed EAST in October 2003 so that it is easy for domestic and international experts of pronounce and memorize and at the same time more convenient for precise conveyance of its scientific significance.

EAST originates from the initial of "Experimental", "Advanced", "Superconducting" and "tokamak". In Chinese, it means the same of the four English words. Besides, as EAST carries the meaning of "east", its Chinese name is "Dongfangchaohuan" (super circulator in the east) (Fig. 14).

EAST project was launched in 1998 and carried out in 2000. After five years in 2005, final assembly of the facility was completed. Its design, R&D, processing and assembly were all accomplished by researchers and technicians of IPP. After completion, the main vessel of the EAST facility stands 11 m tall and 8 m wide and weighs 400 tons. The facility is composed of six main components, i.e. the super high vacuum vessel, toroidal fields, poloidal fields, internal and external cold shield, outer vacuum cryostat and supporting system. As the first fully superconducting non-circular cross-section tokamak, EAST adopts superconducting materials for all 30 inner fields (including 16 toroidal fields and 14 poloidal fields) (Fig. 15).

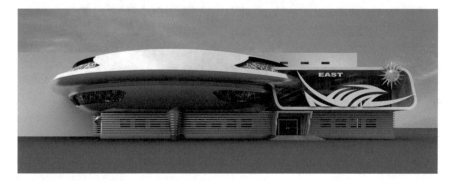

Fig. 13 Design of the EAST control hall

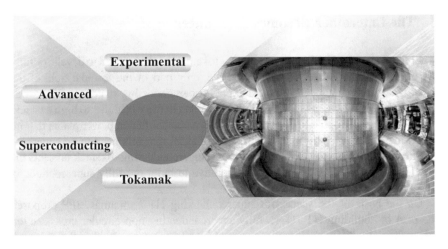

Fig. 14 EAST is an acronym of four English words

Fig. 15 Main vessel of the fully superconducting EAST

In 2006, engineering commissioning of EAST was completed and the first discharge was also achieved. Successful development and operation of EAST has laid a solid foundation in physics, engineering technologies and talent cultivation for China to continue the magnetic confinement fusion research. The EAST International Advisory Committee said after the on-site inspection that "this brilliant achievement

Fig. 16 The inner vacuum chamber of EAST

remains a critical milestone in the development history of nuclear fusion energy across the world" (Fig. 16).

In 2012, EAST achieved sustained discharge of high-temperature plasmas of 20 million degrees Celsius for 411 s. This is the longest record around the globe in terms of the continuous operation of the tokamak fusion facility. In July 2007, EAST achieved long-pulse high-confinement plasma operation stably for 101.2 s. This makes the facility the first tokamak experimental facility in the world to achieve stable high-confinement operation at the hundred-second level. These data drive forward humanity's progress in nuclear fusion field and have received high recognition from the international nuclear fusion community (Fig. 17).

4 Development with International Cooperation

Energy crisis and environmental issues are the common challenges of humanity. In order to accelerate the process of nuclear fusion research, some countries and regions decided to cooperate to research and develop a larger and more advanced tokamak facility. ITER project came about under such background.

ITER project originates from the 1985 Geneva Summit during which Gorbachev and Regan made a proposal. According to the proposal, the US, the Soviet Union, Europe and Japan began the design of ITER in 1987, completed the conceptual design in 1990 and finished engineering design and some technological pre-research. China joined ITER officially on Feburary 18th 2003, responsible for some design and development tasks of the ITER facility. On November 21st 2006, the EU, Japan,

a China set to make fusion history

The world's first fully superconducting tokamak is soon to produce a discharge of ionized gas or plasma.

If all goes as planned, China's Experimental Advanced Superconducting Tokamak (EAST) project will make its first plasma in the next few weeks.

EAST uses superconducting coils to create a magnetic field that confines plasma inside a doughnut-shaped vessel known as a tokamak. The behaviour of the plasma should shed light on the potential of nuclear fusion as an energy source.

Conventional experimental fusion machines use copper coils, or a combination of copper and superconducting coils, to trap the hot plasma. But copper coils heat up and need to be cooled down regularly, thus limiting operating time. EAST has only superconducting coils so it can be operated continuously.

The US$25-million machine sets the stage for the multibillion-dollar ITER fusion experiment that is to be built in France; ITER, due to start operations in 2016, is similarly designed to be all-superconducting.

"We'll need new energy resources for a long-term period, and fusion will be one of them," says Peide Weng,

deputy manager and chief engineer of the EAST project at the Institute of Plasma Physics of the Chinese Academy of Sciences. "For commercial use, it should be superconducting because it will need continuous operation."

China approved the machine in 1998, as part of a push towards new energy sources. Construction then began in 2000 in Hefei, in southern China. The 150-member EAST team imported some material and components, but designed and fabricated the bulk of the equipment on its own.

EAST is only one-tenth the volume of Japan's JT-60 tokamak, and one-hundredth the expected volume of ITER. It won't produce fusion power, and is designed to study advanced tokamak physics. The first plasma, created from heated hydrogen gas, will probably last for only a few seconds. Still, "it will be a very important step forward," says Toshihide Tsunematsu, director-general of the Naka Fusion Institute of the Japan Atomic Energy Agency, who visited EAST a few weeks ago. The agency owns the JT-60 tokamak.

Eventually, the EAST team aims to hold a plasma for study for as long as 1,000 seconds. In other tokamaks plasmas last

for only a few tens of seconds.

South Korea is currently developing a tokamak similar to EAST, called the Korean Superconducting Tokamak Reactor (KSTAR), whose construction is expected to be completed at the end of 2007. Japan also plans to upgrade its JT-60 machine to make it fully superconducting in a few years.

International physicists praise what China has accomplished so far. In 2003, 25 physicists visited EAST as part of its international advisory committee.

"Everybody came away very impressed," says Dale Meade, a physicist with the Princeton Plasma Physics Laboratory in Princeton, New Jersey, and a member of the group. The committee plans to hold another meeting in October, when China hosts a conference of the International Atomic Energy Agency.

In the meantime, the EAST researchers have plenty to work on, says Tsunematsu. They will have to improve key technologies, such as a device to heat the plasma, and be able to effectively control high-temperature plasma for a long period of time. "China will face a real challenge," he says. ∎
Ichiko Fuyuno
With additional reporting by Geoff Brumfiel

In the lead: if China's EAST project is a success, it will pave the way for other major fusion experiments around the world.

Fig. 17 The coverage of EAST in *Science* and *Nature*

b

Speed matters. It has taken just over 5 years and $63.2 million to complete China's new tokamak, according to the Institute of Plasma Physics.

ENERGY ALTERNATIVES

Waiting for ITER, Fusion Jocks Look EAST

China is breaking new ground with a fusion test bed that will tide researchers over until the ITER megaproject comes online

HEFEI, CHINA—The official launch of the International Thermonuclear Experimental Reactor (ITER) project next week will mark a coming of age for fusion research in Asia. When the $11 billion effort was initiated in 1985, ITER's four original backers—the United States, the European Union, Japan, and the Soviet Union—accounted for nearly all worldwide research into harnessing fusion, the process that powers the sun, to produce energy. But now the three newest ITER partners, China, South Korea, and India, are showing that they didn't just buy their way into one of the biggest physics experiments since the Manhattan Project: They are contributing crucial expertise as well.

The first new Asian fusion tiger out of the gate is the Institute of Plasma Physics (IPP) of the Chinese Academy of Sciences, which in March completed testing a machine that has never been built before: a fully superconducting tokamak. This toroidal vessel isn't the largest or most powerful device for containing the superhot plasma in which hydrogen isotopes fuse and release energy. But until India and South Korea bring similar machines online (see sidebar, p. 993), it will be the only tokamak capable of confining a plasma for up to 1000 seconds, instead of the tens of seconds that machines elsewhere can muster. ITER, expected to be completed in Cadarache, France, in 2016, will have to sustain plasmas far longer to

Fire when ready. EAST will fill a crucial gap for fusion researchers until ITER is built, says Director Wan Yuanxi.

demonstrate fusion as a viable energy source. But researchers from China and around the world will be able to use IPP's Experimental Advanced Superconducting Tokamak (EAST) to get a head start on learning to tame plasmas for extended periods. "This will make a big contribution for the future of fusion reactors," declares Wan Yuanxi, a plasma physicist who heads EAST.

Fusion research over the next decade will be probing the physics of steady-state plasmas like those promised by ITER, says Ronald Stambaugh, vice president for the Magnetic Fusion Energy Program at General Atomics in San Diego, California. "EAST will play a big role in that," he says. Others credit IPP for building its advanced tokamak fast, in just over 5 years, on a shoestring $37 million budget. That's a fraction of what it would have cost in the United States, says Kenneth Gentle, a plasma physicist and director of the Fusion Research Center at the University of Texas, Austin. "That they did this in spite of the financial constraints is an enormous testimony to their will and creativity," adds Richard Hawryluk, deputy director of the Princeton Plasma Physics Laboratory.

IPP adroitly fills a generational gap. Fusion power will rely on heating hydrogen isotopes to more than 100 million degrees Celsius, until they fuse into heavier nuclei. The leading device for containing this fireball is the tokamak, a doughnut-shaped vacuum chamber in which a spiraling magnetic field confines the plasma. Ringlike metal coils spaced around the doughnut—toroidal field coils—and a current in the plasma produce this spiraling field. Additional coils in the center of the doughnut and along its circumference—poloidal field coils—induce the current in the plasma and control its shape and position.

Early tokamaks had circular cross sections and copper coils, which can only operate at peak power in brief pulses before overheating. ITER will be far more sophisticated. It will have a D-shaped cross section, designed to create a denser plasma that can generate its own current to supplement the induced current, reducing energy input. And coils will be superconducting. (No major tokamak has had superconducting poloidal field coils.) At temperatures approaching absolute zero, superconductors carry current without generating resistance, allowing more powerful magnetic fields that can be maintained far longer.

Researchers want to try out a D-shaped, fully superconducting test bed before scaling up to ITER, which will be two to three times the size of current tokamaks. The Princeton Plasma Physics Laboratory had planned to build such a device. But a cost-conscious U.S. Congress killed their $750 million Tokamak Physics Experiment in 1995. EAST and the two other Asian tokamaks under construction intend to fill this gap.

"We recognized this was an opportunity for us to make a contribution for fusion research," Wan says. For support, he tapped into China's worries about its growing demand for energy. "There is no way we can rely entirely on fossil fuels," he says. China's government approved EAST in 1998.

IPP faced an enormous challenge. The institute, founded in 1978, had built a few tiny tokamaks in the 1980s and got a hand-me-down, partially superconducting tokamak from Russia's Kurchatov Institute in 1991. EAST would be a totally different beast. "We didn't have any experience in the design, fabrication, or assembly of these kinds of magnets," Wan admits. Neither did Chinese manufacturers.

Industrial partners supplied parts of the tokamak, including the vacuum vessel. But the superconducting coils and many other high-tech components would have been too expensive to import. "We had to do [these] ourselves," says the tokamak's chief engineer, Wu Songtao. So Wu's team bought precision milling machines, fabricated their own coil winders, and built a facility to test materials and components at cryogenic temperatures. "They literally built a whole manufacturing facility on site," says Hawryluk.

19 MAY 2006 VOL 312 **SCIENCE** www.sciencemag.org
Published by AAAS

Fig. 17 (continued)

Russia, China, South Korea, India and the US signed an official agreement and chose Cadarache in the south of France as the site for ITER (Figs. 18 and 19).

ITER is the largest fully superconducting tokamak fusion reactor in the world. It stands 29 m tall and 28 m wide and weighs 23,000 tons. The overall investment totals 10 billion euros calculated according to the price in 1998. Meanwhile,

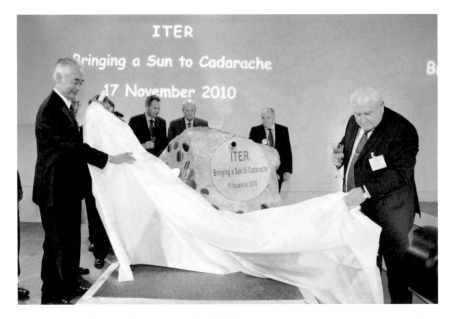

Fig. 18 Cadarache in south France was the ITER site

Fig. 19 ITER facility construction site

ITER facility is applied with advanced technologies. It gathers latest techniques and technologies by that time. Its design is based on the research and progress plan jointly made by research groups and businesses of the field. It is no exaggeration to say that the current nuclear confinement research has entered the ITER era. ITER project will be completed in 2025 as scheduled at present. Due to long duration of R&D, investment for ITER has been increasing. According to estimates, the investment has surpassed that of the International Space Station, making ITER the biggest technological cooperation project (Fig. 20).

As one of the seven ITER partners, China undertakes about 9% of the research and development task and enjoys over 5% of the human resourceallocation right. The percentage is increasing as ITER project proceeds and China's strength in science and technology improves.

As one of China's bases in controlled nuclear fusion research, IPP undertakes over 70% of the research tasks China responsible for in ITER project. Thetasks involve design of the power supply system for the ITER poloidal field converter

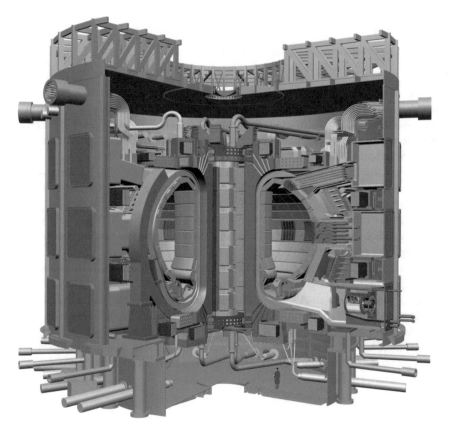

Fig. 20 Main vessel design of ITER

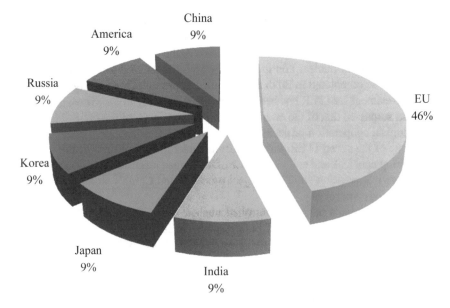

Fig. 21 Contribution of member states of ITER project

and domestic integration, research on the ITER large-scale superconducting magnet feeder system, coil manufacturing project for the ITER superconducting alignment filed, Nb3Sn superconducting coil R&D project for ITER superconducting magnetic system and the ITER PF6 superconducting coil manufacturing project obtained in an independent bidding (Fig. 21).

As the tokamak facility "closest" to ITER, EAST is designed with similar magnetic configuration of advanced non-circular cross-section divertor, poloidal system in integrated design, radiofrequency-wave heating etc. These features make EAST a platform for testing key scientific and technological problems and hugely reduce risks of ITER. Based on EAST, China will make significant contribution to ITER development and operation (Fig. 22).

As a large-scale superconducting tokamak under construction, ITER is aimed at realizing "plasma combustion" and verifying the scientific and technological feasibility of nuclear fusion. Although there is still a long way to go for commercial application of ITER, humanity have already seen the silver lining in peaceful use of nuclear fusion energy. Full participation in ITER project is a challenge and an opportunity for China. The project will facilitate "going global" and "introduction" of relevant nuclear fusion research achievements to achieve leap-forward development of the research in China.

Fig. 22 IPP signed the first contract for ITER procurement package in 2009

5 Fusion Reactor Development Plan of China

Although ITER is still in construction, many countries have formulated nuclear fusion development roadmaps. According to China's plan, the country will welcome the new-generation Fusion Engineering Test Reactor (CFETR). This facility is targeted at, before building the nuclear fusion power plant, designing and developing a fusion engineering test reactor based on China's experience in design and operation of multiple tokamaks and the ITER project (Fig. 23).

Knowledge Link
CFETR

   CFETR has been designed since 2011. It is aimed at engineering conceptual design of China's fusion engineering test reactor, integrated design and R&D of key technologies for the fusion reactor based on key advanced technologies in magnetic confinement fusion. CFETR has strategic significance for China in terms of complete master of technologies for designing and developing the next-generation fusion reactor.

   As of now, CFETR has been set scientific and engineering objectives and the overall arrangement and key parameters have also been confirmed. On the basis of calculation and engineering feasibility verification, detailed engineering conceptual

Fig. 23 Design of CFETR building complex

design of the main superconducting vessel has been completed. In addition, initial conceptual design of the tritium plant and power supply system etc. has also been accomplished. Some major research projects have been launched with some progress already achieved (Fig. 24).

CFETR will fill the scientific and technological gap between ITER and DEMO and demonstrate engineering feasibility of continuous large-scale nuclear fusion energy for safe and stable power generation. It is therefore called "engineering test reactor". Compared with ITER, CFETR is advanced in two aspects: one is achieving stable state or long-pulsed plasma combustion and realizing 30–50% (4–5% for ITER) of effective operation time during its life cycle; the other is achieving self-sustainability of tritium inside the cladding, or realizing self supply of tritium. The second is a step forward towards commercial use compared with ITER which requires manual supply of tritium.

While participating in ITER project, China also plans for its own fusion engineering test reactor. This arrangement helps not only undermine effects of ITER delay on the country's nuclear fusion research, but also learn thoroughly key technologies of ITER so as to master relevant physical and engineering technologies in fusion reactor.

6 Outlook on China's Road to Nuclear Fusion

From CT-6 to EAST, nearly a half century has passed by. In this journey of magnetic confinement fusion, generations of researchers in China have solved numerous physical and engineering problems. They started from scratch, developed from the weak to the strong, from a follower to a forerunner for leap-forward development. Back

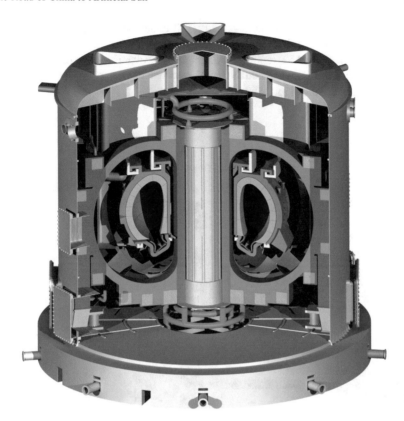

Fig. 24 Design of the CFETR main vessel

to the past, successful development and operation of generations of tokamak facilities have laid a sound engineering and physical foundation for stable and advanced fusion test reactors in the future. Looking into the future, nuclear fusion researchers in China have drawn a comprehensive and visionary roadmap that draws on fusion engineering test reactor and incorporates DEMO test objectives and targets nuclear fusion power plants.

Currently, two prior projects of CFETR—National Applied Superconducting Engineering Technology Centerand Divertor Test Platform have been incorporated in the Major National Scientific and Technological Infrastructure Program under the "13th Five-year Plan". Pre-research and prior design have been conducted before the official approval is given by relevant authority. After the approval is obtained, China will continue to lead magnetic confinement research towards higher scientific objectives and peaceful development and utilization of the energy.

Energy crisis is a common topic discussed by the whole world and peaceful use of magnetic confinement fusion remains a common dream of humanity. We sincerely hope that the first light of the artificial sun comes from China just as the first fully superconducting tokamak in the world was born in China.

Now, we believe that you have gained a full understanding of China's development process of the artificial sun. As the first non-circular cross-section tokamak
facility in the world, EAST represents the highest level and achievement of China's
tokamak. Next, we will look inside the EAST tokamak facility to see what the extreme
environments there are (Fig. 25).

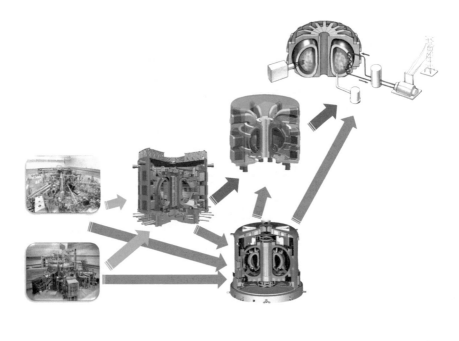

Fig. 25 Roadmap of China's magnetic confinement fusion development

How Cool is Artificial Sun

Jiao Peng

Abstract In order to make controlled nuclear fusion as stable as the sun, extreme environments are set inside the artificial sun, including ultra-high and ultra-low temperatures, ultra-high vacuum, ultra-strong magnetic fields and ultra high-voltage currents. This chapter brings an inside view of the tokamak.

1 A Scorching Heart with Ultra-High Temperature

The objective of tokamak is to realize controlled nuclear fusion. When it comes to extreme environments inside the facility, the primary focus that deserves attention is the scorching "heart". In order to create conditions for fusion, the core temperature of plasmas in the artificial sun needs to reach, generally speaking, 100 million °C or even higher. This is much higher than what we've imagined as the core temperature of the sun stands only at 15 million °C—much lower compared to the artificial sun (Fig. 1).

Only under ultra-high temperature, fusion materials—nuclei of deuterium and tritium will move at the speed high enough to overcome the huge repulsion existing between each nucleus. When such condition is realized, D-T nuclei will approach close enough to the degree that nuclear force can exert an influence. Two nuclei will then combine to achieve fusion reactions and release huge energy at the same time. The first step for smooth fusion reactions is to heat the materials and maintain ultra-high temperature (Fig. 2).

Of course, with such high temperature, nuclear fusion fuels will become plasmas. What's more, any material will turn into plasmas under such ultra-high temperature. These ultra high-temperature plasmas will try to "escape" all around like a wild horse. Tokamak has to be installed, therefore, with a "heart" strong enough to bear these plasmas.

J. Peng (✉)

Institute of Plasma Physics, Chinese Academy of Sciences, Hefei, Anhui, China

e-mail: yihco0101@sina.com

© Zhejiang Education Publishing House 2021

B. Wan (ed.), *Man-Made Sun*, China's Big Science Facilities,

https://doi.org/10.1007/978-981-16-3887-9_7

Fig. 1 Scorching core

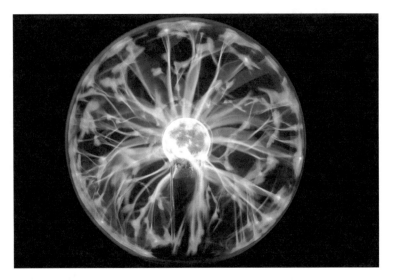

Fig. 2 Nuclear fusion materials under high temperature will turn into plasmas

2 A Heart with Ultra-Strong Magnetic Fields

Scorching temperature requires a strong "heart". For tokamak facility, the strong heart is the ultra-strong magnetic field. For example, filed strength at the plasma center reaches 3.5 Telsa and the maximum strength of ITER can reach 5.3 Telsa, compared to about 0.00005–0.00006 Telsa of the magnetic strength of the earth.

Therefore, the field strength at the plasma center of the tokamak facility is 100,000 folds of the magnetic field strength of the earth. Actually, magnetic fields inside the tokamak fall into two categories, i.e. the changing field and the stable field. They are both strong and yet different in functions.

Strong changing magnetic fields will constantly turn a large volume of power into inner energy of fusion materials which pushes up plasma temperature. Therefore, the primary task of the changing field is to produce and heat plasmas. However, the heating also faces limitation in the field strength (about 10 million °C at most). Additional heating is thus required for tokamak system.

Besides, the changing field can also adjust the configuration and location of plasmas as needed. The stable strong field is like invisible toroidal chains that firmly "tie" fast-moving plasmas and only allow for "chain"-centered movement. During this process, ultra-high temperature plasma confinement is achieved to prevent materials in the tokamak from being burnt down by the ultra high-temperature plasmas.

Knowledge Link
The magnetic system of EAST
   The magnetic system of EAST includes 30 fields, of which 14 are poloidal fields (PF) and 16 are toroidal fields (TF). The PF is composed of 6 central solenoid fields (PF1–PF6) and 8 external large fields (PF7–PF14) at the core. Changing fields are all induced by changing currents placed horizontally.

Changing fields generated by the central solenoid fields are used to stimulate and heat plasmas. Changing fields produced by large fields are employed to control configuration and location of plasmas. 16 D-shaped toroidal fields are placed vertically to produce toroidal stable fields to confine plasmas (Fig. 3).

The magnetic system is one of the most critical components of the fully superconducting tokamak. Its overall weight exceeds one fourth of the entire facility and the same proportion works in its overall cost. Of course, a large magnetic system is not enough. The key is that an ultra-large volume of power needs to be connected to make the magnet generate hugely powerful magnetic field.

3 "Blood-Surging" Electric Currents

If the magnetic system is compared to the heart which generated ultra-strong magnetic field, the ultra-large volume of electric currents is like fresh blood passing through the heart. The maximum current volume in poloidal fields of EAST reaches 15,000 A, and the figure for toroidal fields is 16,000 A.

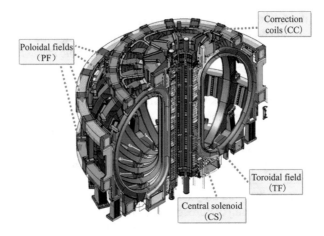

Fig. 3 Tokamak magnetic system

As we know, ordinary conductors have resistance. When electric currents pass through, the conductor will consume energy which produces heat. For this reason, conductors generally need to be thick enough to avoid being burned.

Home appliances generally consume electricity no more than 10 A, in which case an ordinary conductor is enough. Yet a thicker conductor will be required in case of some appliances that require large volume of electricity (16 A). The maximum volume of electric currents for EAST can be as high as 16,000 A, or 1000 folds of home appliances. If an ordinary conductor is still employed, we need to make it as thick as a common pillar so as to prevent it from being burnt down and to save energy (Fig. 4).

If the conductor is thickened, there will be some problems. First of all, a thick conductor means huge difficulty in deciding the size of the nuclear fusion facility and in constructing the facility. Second, because energy will generate a large volume of heat, there will be risk for the conductor to be burnt down if the heat is not released timely. Finally, as much input energy is consumed in the conductor, there will be huge demand for external supplement, which poses huge challenges in realizing the objective of "output exceeding input". The magnetic system of the EAST facility has been installed with ultra low-temperature cooling superconductors, which will ensure that thin conductors can also bear enormous electricity (Fig. 5).

Knowledge Link

Power supply system of EAST

Magnetic currents in the EAST facility fall into two types. One is the stable ultra-huge current moving in toroidal fields which generate ultra-strong doughnut-like magnetic fields; the other is the fast-changing large pulsed

Fig. 4 There is resistance inside an ordinary conductor

current moving in poloidal fields. The second is the supporting currents to stabilize the magnetic field.

Meanwhile, a superconducting magnetic feeder system is needed to ensure safe and effective movement of these currents. This system will connect the power supply system that generates currents with superconducting magnets where the currents move. This superconducting feeder system achieves not just current connection and energy transmission, but also transition and insulation between different temperature zones.

4 Ultra-Low Temperature Beneath the "Cool" Surface

For the EAST facility, the superconductor that carries an ultra-large volume of currents is located one meter away from the ultra high-temperature core. To ensure stable superconducting state, the superconductor needs to be immersed in liquid helium of − 269 °C (Fig. 6).

That is to say, a temperature gap from 100 million to − 269 °C exists within one-meter distance from the core of The EAST facility. This shows the huge difficulty in developing the artificial sun. − 269 °C is the temperature of the supercritical liquid

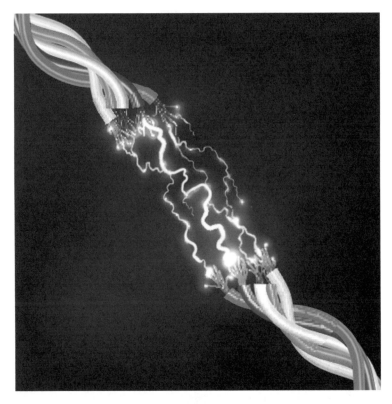

Fig. 5 Ordinary conductors cannot bear a huge amount of currents of the EAST facility

Fig. 6 The superconductor of the EAST facility requires low temperature to maintain the superconducting state

Fig. 7 Magnets in superconducting state

helium, while the lowest temperature record of the earth is about $- 90\ ^\circ\mathrm{C}$, close to the minimum temperature in Physics ($- 273.15\ ^\circ\mathrm{C}$). At such a low temperature, superconductors will show very peculiar superconducting characteristics (Fig. 7).

Superconductivity is a physical phenomenon, i.e. the resistance of some special materials will disappear at low temperature. Without resistance, heat will not be generated no matter how strong the current is. In this case, a very thin conductor will be strong enough to bear very strong currents. This means that the superconducting magnet has smaller volume, stronger magnetic field and longer operation time. Superconductors have therefore been very widely applied to tokamaks that carry very strong currents.

Superconducting tokamaks basically employ cryogenic superconducting materials, such as EAST with NbTi materials. In the future, nuclear fusion facilities will also adopt Niobium-tin (Nb_3Sn) materials which enable higher critical magnetic fields and higher critical currents. In order to maintain superconductivity of the material, these superconductors all need to work at temperature of $- 269\ ^\circ\mathrm{C}$. Yet the price is that it will be more difficult to maintain the cooling system and more complicated to develop the facility.

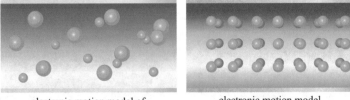

electronic motion model of
conventional conductor

electronic motion model
of superconductor

Fig. 8 Models of electron movement in the conventional conductor and the superconductor

Knowledge Link
Superconducting materials
   Depending on different working temperatures, superconductors are usually categorized into low-temperature superconductors and high-temperature ones. This is a relative concept and the differentiation is not so clear. In general, high-temperature superconductors refer to those working at − 230 °C or even lower temperatures. Although some high-temperature superconductors have higher critical magnetic fields and more powerful critical currents at lower cooling costs than low-temperature ones, high-temperature superconductors for the large-scale fusion superconducting magnetic system are still in research stage due to their physical constraints and special requirements for magnetic systems.

At present, IPP of CAS is developing the large-scale superconducting magnet based on the high-temperature superconducting material Bi-2212. If the development succeeds, the magnet will be primarily applied to central solenoid magnets of CFETR (Fig. 8).

The EAST facility has a burning heart and a cool appearance. Just as fire and ice dance together, these two extreme temperatures are harmonized in the main vessel with a diameter of 8 m, and the shortest distance stands only at 1.2 m. The question then arises: how does such a wide temperature gap exist in such a short distance? The answer comes from the unrivalled insulation design of the EAST facility.

5   A Purified Heart with Ultra-High Vacuum

In the EAST facility, a special insulation design is adopted in order to insulate over 100 million °C of ultra high temperature and − 269 °C of ultra low temperature. That is vacuum.

Vacuum is a very thin gas state with very low particle density and quite slow heat conduction. It thus remains an ideal option for insulation. In our daily life, the thermos bottle follows the principle of vacuum insulation. While the vacuum degree of the thermos bottle is low, the demand for superconducting tokamaks such as EAST is much higher and the vacuum chamber is also much larger. The core vacuum of EAST can be maintained at 10^{-6} Pa, which is equivalent to 1/100 billion of atmospheric pressure we normally feel.

Such high vacuum for insulation realizes coexistence of the two extreme temperatures achieved within a very short distance. At the same time, the vacuum protects materials of the facility from being burnt down by the fusion fuel at the temperature of over 100 million °C.

The EAST facility is comprised of the inner vacuum chamber and the outer chamber. The later is used for insulation and the former is used, in addition to insulation, mainly as the vessel for nuclear fusion reactions. The inner vacuum chamber is therefore one of the key components of the EAST facility which accounts for around one third of the overall weight.

Apart from the inner and outer vacuum chambers, EAST also needs a high level of vacuum for many of its external systems. To develop and maintain such vacuum requires coordination of mechanical pump, Rhodes pump, molecular pump, cryogenic pump etc.

Now you may have understood how cool the artificial sun is. The extreme environments—ultra-high temperature, ultra-low temperature, ultra-large currents, ultra-strong magnetic field and ultra-high vacuum—that transcend our cognition, turn the artificial sun an entity of "ultra" problems. These environments are the manifestation of the harshness in design and development of a tokamak facility (Figs. 9 and 10).

Although scientists around the world have solved many problems and made huge progress and desirable achievements, the road to the artificial sun is till long and harsh. What is the destination of this road in the future? Will the artificial sun be developed? The next chapter will give you answers.

Fig. 9 The thermos bottle follows the vacuum principle for thermal insulation

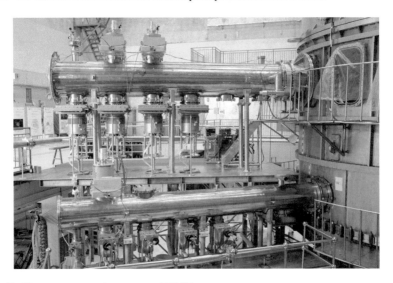

Fig. 10 The vacuum pumping system of EAST

Artificial Sun Is Not Just a Dream

Rong Yan

Abstract With hard work over the past half century, the world has witnessed successes in development and operation of multiple tokamaks and major achievements in magnetic fusion confinement. We believe that the success to the brilliant artificial sun is within reach as nuclear fusion research proceeds to the era of energy development.

1 A Thousand-Mile Journey Begins with a Single Step

In the journey of the artificial sun research over half century, countries all over the world have built over 100 tokamak facilities. Some breakthroughs are mainly achieved in large-scale tokamak facilities. In the initial stage of the tokamak research, the discharge only lasted for several seconds. Thanks to continuous hard work, the duration was prolonged gradually from seconds, dozens of seconds to several hundreds of seconds today. Humanity has made tremendous progress in fusion research compared with the situation during the early days (Figs. 1 and 2).

However, we still have a long way to go before realizing fusion power. Currently, controlled fusion research platforms developed by countries in the world can only be regarded as nuclear fusion experimental facilities. They are basically targeted at physical mechanisms and engineering technologies and thus not yet involved with energy output after the fusion reaction. Our ultimate objective is to realize controlled fusion reactions and builds commercial nuclear fusion power plants to output the energy generated in the reaction. Therefore, there still remain many major problems to be solved on the road to successful commercial fusion power plants from the current tokamak experimental facilities (Fig. 3).

However, it is still costly to build such fusion experimental facility even without energy output units. Every facility costs from several hundred millions or even

R. Yan (✉)
Institute of Plasma Physics, Chinese Academy of Sciences, Hefei, Anhui, China
e-mail: yihco0101@sina.com

© Zhejiang Education Publishing House 2021
B. Wan (ed.), *Man-Made Sun*, China's Big Science Facilities,
https://doi.org/10.1007/978-981-16-3887-9_8

Fig. 1 Humanity have continuously progressed in nuclear fusion research

Fig. 2 Nuclear fusion research still confronts many obstacles

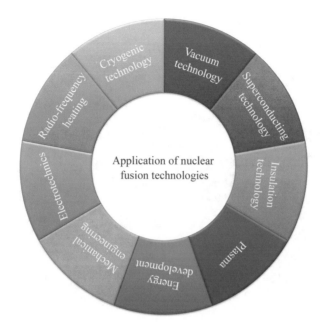

Fig. 3 Nuclear fusion research involves many advanced technologies

billions. These facilities are developed with advanced technologies, including large-scale superconducting magnets, high-volume ultra-high vacuum, long-pulsed high-power plasma acceleration and radio frequency technology, plasma control engineering etc. As the research goes further and the facility is enlarged, the requirement for plasma engineering technologies and diagnosis will be higher. Scientists are decisive in the spirit that "a journey of a thousand miles begins with a single step" to lay a solid foundation in every step. They begin with developing fusion experimental facilities to fully understand fusion control principles and master various advanced engineering technologies to have a full technological preparation. They then move on to build the next generation of fusion reactors and the future demonstration fusion reactors. The destination is successful commercial fusion power plants which enables stable output of the energy generated from fusion reactions to support people's life and production activities in the future.

2 Courageous Steps to the Future

With the rapid development of tokamak experimental facilities, the fusion community gradually pays attention to research on engineering technologies of fusion reactors. Since the late 1970s, researchers have conducted considerable research on fusion

Fig. 4 ITER project remains the largest international cooperation project

reactor design, key materials, tritium technology, remote control and operation technologies etc. Till this stage, the community believes that enough knowledge and technologies have been acquired to build a tokamak test reactor that enables ignition. On the basis of this consensus, an unprecedented international cooperation project—ITER comes about (Fig. 4).

The ITER project has been through several twists and turns from the beginning stage till now. The number of partners has increased from four to seven; the investment has been rising based on the initial scheme of 5 billion euros + 5 billion euros (development + operation), making it the largest technological cooperation project in the world. Yet undoubtedly, much progress has been made to promote the development of fusion energy. From another perspective, ITER also sets an example for win–win cooperation compared to isolation before.

ITER is currently under development and is estimated to produce the first plasmas in 2025. The next two decades following 2025 will be the test operation stage of plasmas. This stage is comprised of two steps: the first is to explore pulsed plasma movement driven by the superconducting magnets and pursue high fusion energy gain factor—$Q > 10$ (500 s for reaction); the second step is targeted at long-pulsed stable movement and to achieve more than 3,000 s ($Q > 5$) of sustained plasma combustion duration (Fig. 5).

Fig. 5 ITER facility under development

Another important technological task of ITER is to test performance of the components under nuclear fusion environment. This includes detection of irradiation defects, high heat load, and heavy electro dynamic shock, real-time and local development of large-scale tritium technologies. These must be completed only with ITER and before the design and development of commercial reactors. The mainstream opinion on the ITER project is that ITER stands ready in the science and engineering technologies for the successful development and operation; after the stage of the test reactor and the prototype reactor power plant, fusion energy for commercial use is expected to be realized by the mid twenty-first century.

Knowledge Link

The history of nuclear fusion

Over six decades have passed for humanity to develop nuclear fusion energy from the initial simple and small tokamak experimental facilities, to the current large-scale full superconducting tokamak experimental facility and to the development of fusion test reactors. We should be fully aware of the daunting challenges in fusion energy development. At the first meeting of the Geneva Conference in 1955, the Indian scientist Bhabha believed that nuclear fusion would be realized in 20 years, an optimistic estimation which many renowned scientists at the meeting agreed. In 1957, the Soviet scientist Kurchatov shared his fully optimistic outlook on the fusion research when he introduced nuclear fusion achievement of the Soviet Union for the first time at the Harwell research center in the UK. The reality, however, shows their underestimated complexity of the research. A long and challenging road still lies ahead in the application of the nuclear fusion energy in the future.

Although it is a concrete step forwards for several major countries to jointly develop ITER reactors, ITER is essentially a test reactor and many problems still need to be solved—such as the circular and self-sustained state of tritium and fusion energy output—before the prototype reactor comes out. Under such background, many countries propose to build DEMO reactors so as to address these engineering problems.

China proposes the CFETR project which will proceed in two stages: the first stage is to learn and master relevant technologies and address engineering problems unsolved in the ITER project; the second is to, after upgrading the facility, master and improve engineering technologies required for building the demo commercial fusion reactor. At present, researchers in China focus on pre-research of key CFETR components and detailed engineering design will come out soon.

3 Triumphant Fanfare Is On

The fusion community proposed the concept of DEMO after ITER and before the prototype of commercial fusion reactors. DEMO is the key to realizing commercial fusion energy. Major technologies addressed by DEMO include fusion energy output, reliable and practical self-sustainability of tritium (Fig. 6).

Members of the ITER project considered various types of DEMO. The options are to be confirmed from technological and physical integration perspectives. One that deserves attention is an ambitious project proposed by the EU. Its main objective is to design a very sophisticated DEMO and the key strategy is to develop two conceptual DEMOs at the same time, i.e. pulsed tokamak DEMO1 and stable tokamak DEMO2.

China has taken a step forward. Up till now, China has finished overall engineering conceptual design of CFETR and is conducting pre-research of its key components. The detailed engineering design is also about to began. During the late period of the 13th Five-Year Plan, China will independently build the fusion engineering test reactor of 200,000 to 1 million kW and CFETR is expected to be completed in around 2030. The main objective of the second stage is to master and improve engineering technologies required for building the commercial fusion DEMO reactor (Figs. 7 and 8).

The development of CFETR will not only lay a solid foundation in science and technology and engineering for independent development and peaceful use of fusion energy, but also make it possible for China to be the first to generate power with nuclear fusion energy and achieve leap-forward development.

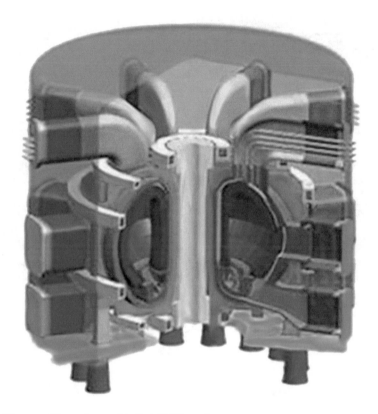

Fig. 6 Diagram of the main vessel in the DEMO reactor

4 The Dream Finally Comes True

Humanity will gradually gain the capacity for building and operating commercial fusion power plants by exploring controlled fusion reactions with existing tokamak facilities and carrying out research on fusion power plants installed with the next generation of test reactors and DEMO reactors. Commercial fusion power plants work in the same way as the current nuclear power plants in energy output. Neutrons with energy heat water to generate steam; the steam then drives the steam engine to generate power. The principle is similar to that of the coal-fueled power plants. Major powers and economies in the world are fully clear of their controlled fusion development plans. The first prototype fusion power plant is expected to be completed in 2050 and the fusion power will reach 1,000–3,000 MW (Fig. 9).

If the first prototype fusion power plant is developed and operated successfully, controlled fusion energy will embrace the commercial era. This will be another energy revolution and will once again change the landscape of the world. According to optimistic estimation, the future energy consumption will undergo following changes:

Fig. 7 Design of CFETR

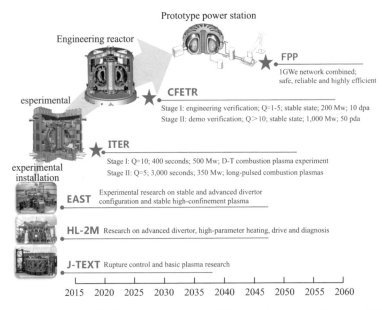

Fig. 8 Roadmap of controlled fusion development in China & mission and objectives of CFETR

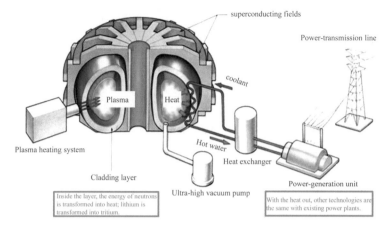

Fig. 9 The principle diagram of the commercial fusion power plant

by 2040, oil consumption will reach its peak; by 2050, bio energy, water energy, geothermal energy, wind energy and solar energy will account for a far large proportion of the energy mix and one third of the total energy consumption, and thermonuclear energy will reach a quarter of the total consumption; by 2100, oil consumption will account for less than 5% of the total energy consumption. Meanwhile, as nuclear fusion energy gradually replaces conventional fossil fuels and nuclear fission energy, humanity will be free from worries of environmental pollution, ecological degradation and conventional issues such as nuclear leakage and nuclear radiation. By that time, the world will embrace clear air and water and tranquil environment. Let us look forward to the coming of the day (Fig. 10).

Fig. 10 The dream of the artificial sun will ultimately come true

Historic Events of Tokamak Nuclear Fusion Confinement Development in China

In 1974, China built the first iron-core-based tokamak facility—CT-6

In 1978, CAS approved the founding of Institute of Plasma Physics.

In 1980, IPP of CAS built the HT-6A facility.

In 1982, HT-6A was upgraded and renamed HT-6B to continue the research.

In 1984, IPP of CAS built the HT-6M facility.

In 1990, IPP of CAS accepted the T-7 tokamak facility given by the Soviet Union.

In 1994, IPP of CAS completed the transformation of T-7 and renamed it HT-7. The institute also achieved the first commissioning of the.

In 1997, the Leading Group of the State Council for Science and Technology Policy approved the HT-7U Big Science Facility Project.

In 2003, China joined the ITER project officially. HT-7U was renamed EAST.

In 2006, the EAST facility was completed and put into operation.

In 2008, HT-7 achieved continuous and repetitive high-temperature plasma discharge for 400 s.

In 2011, the overall engineering conceptual design of China Fusion Engineering Test Reactor (CFETR) officially began.

In 2012, EAST achieved stable high-confinement plasmas for 32 s, a new record in the world.

EAST achieved high-temperature plasma discharge for 411 s, a new record—the longest duration—in the world at that time.

In 2013, HT-7 superconducting tokamak facility "retired".

In 2016, EAST achieved stable high-confinement plasmas for 60 s. EAST is the first tokamak nuclear fusion experimental facility to achieve stable high-confinement operation at minute level in the world.

In 2017, EAST achieved 101.2 s of stable high-confinement plasmas, a world record. EAST became the first tokamak nuclear fusion experimental facility to achieve stable high-confinement operation at hundred-second.

© Zhejiang Education Publishing House 2021   147
B. Wan (ed.), *Man-Made Sun*, China's Big Science Facilities,
https://doi.org/10.1007/978-981-16-3887-9

Printed in the United States
by Baker & Taylor Publisher Services